Lecture Notes in Mathematics 1630

Editors:
A. Dold, Heidelberg
F. Takens, Groningen

T0222607

Springer
Berlin
Heidelberg
New York
Barcelona
Budapest
Hong Kong
London
Milan
Paris
Santa Clara
Singapore
Tokyo

Daniel Neuenschwander

Probabilities on the Heisenberg Group

Limit Theorems and Brownian Motion

Springer

Author

Daniel Neuenschwander
Université de Lausanne
Ecole des Hautes Etudes Commerciales
Institut de Sciences Actuarielles
CH-1015 Lausanne, Switzerland

and

Universität Bern
Institut für mathematische Statistik
und Versicherungslehre
Sidlerstrasse 5
CH-3012 Bern, Switzerland

Cataloging-Data applied for
Die Deutsche Bibliothek - CIP-Einheitsaufnahme

Neuenschwander, Daniel:
Probabilities on the Heisenberg group : limit theorems and Brownian motion. - Berlin ; Heidelberg ; New York ; Barcelona ; Budapest ; Hong Kong ; London ; Milan ; Paris ; Tokyo : Springer, 1996
(Lecture notes in mathematics ; 1630)
ISBN 3-540-61453-2
NE: GT

Mathematics Subject Classification (1991): 60-02, 60B15, 22E25, 47D06, 60F05, 60F15, 60F17, 60F25, 60G10, 60G17, 60G18, 60G42, 60H05, 60J25, 60J35, 60J45, 60J55, 60J60, 60J65, 62F35

ISSN 0075-8434
ISBN 3-540-61453-2 Springer-Verlag Berlin Heidelberg New York

Typesetting: Camera-ready TEX output by the author
SPIN: 10479780 46/3142-543210 - Printed on acid-free paper

Preface

Probability theory on algebraic and geometric structures such as e.g. topological groups has attracted much interest in the literature during the past decades and is a subject of growing importance. Stimuli which can not be overestimated for the research work which has and is currently been done in the field of probability theory on groups and related structures are the regular Oberwolfach conferences organized by L. Schmetterer, H. Heyer, and A. Mukherjea as well as the recent foundation of the "Journal of Theoretical Probability" also by A. Mukherjea.

In this work we will have, from the probabilistic point of view, a closer look at the so-called Heisenberg group. Its structure reflects the Heisenberg uncertainty principle as non-commutativity of the location and the momentum operator. In a certain sense, it is the simplest non-commutative Lie group, so it is clear that in generalizing classical results of probability theory to the non-commutative situation, one naturally passes by this group. Our aim will be to survey, under the limit theoretic aspect and its relation to Brownian motion, certain results which turned out to be valid on the Heisenberg group but which can not (or not yet) be generalized to the whole class of simply connected nilpotent Lie groups. For this wider framework, we refer (among others) to the forthcoming book of Hazod and Siebert (1995). So our work will to a certain degree be a complement to that book in the sense of some sort of a case study.

The second author of the above-mentioned book in preparation, Eberhard Siebert, untimely passed away in 1993. Without his fundamental contributions, the theory would at any rate not be at that level as it is now. It is one of the modest objectives of our book to underline the importance of Siebert's work in the development of probability theory on (in particular non-commutative) groups.

A word about applications: The Heisenberg group turned out to have many applications not only in mathematics itself (and there even in such remote fields such as combinatorics!), but also in physics (where it in fact comes from) and engineering science (signal theory). Due to the physical ignorance of the author, we have not tried to look for applications of the results presented in this work. The author would be delighted to hear one day about applications outside of "pure" mathematics!

It is my great pleasure to express my most sincere and heartfelt gratitude to my teacher and mentor Professor Henri Carnal for his constant benevolent support; to Professor Wilfried Hazod for his kind hospitality at the University of Dortmund; to Professor

René Schott for his kind hospitality at the University Henri Poincaré Nancy I; to Professor Yuri Stepanovič Hohlov and Professor Gyula Pap for many stimulating discussions; and last but not least to the Ingenieurschule Biel and its director Dr. Fredy Sidler for giving me the opportunity of taking a leave in order to continue my research activities and to begin with this work.

Biel-Bienne, May 1996

Daniel Neuenschwander

Contents

Introduction

From a historical point of view (cf. Heyer (1977), Introduction), the development of probability theory on other structures than Euclidean spaces may be traced back to Daniel Bernoulli, who in his astronomical investigations in 1734 assumed planets to be (uniformly distributed) random points on a sphere. Later on, during the first decades of the 20th century, several mathematicians continued to consider (also non-uniform) probability distributions on the circle and the sphere. We mention the names of Rayleigh, Pearson, Perrin, von Mises, Fisher, and Mardia. Nowadays this direction of research is called statistics of directional data. Von Mises and Lévy (1939) considered probability measures on the torus. The pioneering breakthrough to more general compact groups is due to Kawada and Ito (1940). Bochner, during the late fifties, began to study probabilities on locally compact abelian groups. An early overview of the theory is the book of Grenander (1963). Since then, the field has developed into several directions. One milestone was the paper of Hunt (1956), who considered continuous convolution semigroups on Lie groups. He was able to characterize their infinitesimal generators by an analogue of the classical Lévy-Hinčin formula. It turned out that convolution semigroups were a natural framework for studying limit theorems. The state of the art up to 1977 is most exhaustively described in Heyer's (1977) monograph. We also mention the important work of Stroock, Varadhan (1973) and Feinsilver (1978) concerning stochastic processes on Lie groups.

Since the late seventies it became clear that the simply connected nilpotent Lie groups play a special role in probability theory, in particular where limit theorems are concerned. Among the first papers along these lines are those of Crépel, Raugi (1978) and Raugi (1978) concerning the central limit theorem on simply connected nilpotent Lie groups, and that of Crépel, Roynette (1977), which gives a law of the iterated logarithm for the (three-dimensional) so-called Heisenberg group. The latter is the simplest example of a non-commutative simply connected nilpotent Lie group, in a certain sense even generally the simplest non-commutative non-discrete Lie group from the structural point of view. We will give a detailed description of it later. So we think that in aiming at the non-commutative situation, one must come by this group. In 1982, Hazod (1982) introduced a concept of stability on locally compact groups based on convolution semigroups and one-parameter automorphism groups. Later it was proved by Hazod and Siebert (1986) that strictly stable semigroups are concentrated on the contractible part of the corresponding one-parameter automorphism group, which is isomorphic to a simply connected nilpotent Lie group. On the other hand, Burrell and McCrudden (1974) have shown that any infinitely divisible probability measure on a simply connected nilpotent Lie group is embeddable into a continuous convolution semigroup. The theory of (semi-)stable semigroups on simply connected nilpotent Lie groups is in full growth at present: among the most important ones we mention the papers of Hazod (1982, 1984a, 1984b, 1986), Hazod, Siebert (1986, 1988), Drisch, Gallardo (1984), Nobel (1991), Hazod, Nobel (1989), Hazod, Scheffler (1993), Scheffler (1993, 1994, 1995a, 1995b), Carnal (1986), Kunita (1994a, 1994b, 1995), Neuenschwander, Scheffler (1996), Neuenschwander (1995a, 1995c, 1995d), and the theses of Nobel (1988), Scheffler (1992) and Neuenschwander (1991). Parallel to this stream of research several other aspects of probability theory on simply connected nilpotent Lie groups have been investigated. As some examples we mention further (weak and strong) limit theorems (Pap (1991a, 1991b, 1992, 1993, 1995), Berthuet (1979, 1986), Helmes (1986), Baldi (1986, 1990), Chaleyat-Maurel, Le Gall (1989), Ohring (1993), Neuenschwander (1992, 1995b, 1995e, 1995f), Neuenschwander, Scheffler (1995), Neuenschwander, Schott (1995)), the question of uniqueness of convolution semigroups (Pap (1994)), the explicit construction of Brownian motion with stochastic integrals (Roynette (1975)), as well as the development of a potential theory by Gallardo (1982).

The aim of this work is to give an account of certain limit theorems and of some aspects of Brownian motion under the limit-theoretic point of view on the (three-dimensional) Heisenberg group, with the major part grouped a little bit around the author's own results. The literature being fairly numerous, no claim to completeness is made. Let us also mention the important papers Gaveau (1977), Métivier (1980), and Helffer (1980). In general, only aspects which are (or up to now have been) special for the Heisenberg groups (or for simply connected step 2-nilpotent Lie groups) are taken into consideration in detail. Though the results are often available also for higher-dimensional Heisenberg groups or for all simply connected step 2-nilpotent Lie groups, we do not aim at maximal generality but will restrict ourselves to the simplest case of the three-dimensional Heisenberg group $\mathbb{H} = \mathbb{H}^1$ as a prototype in order to show as simply as possible the ideas and to unify the presentation, for otherwise the text would become too heterogeneous. For the more general formulations we can in most cases refer to the corresponding original works cited. The central component of Brownian motion on $\mathbb{H}$ is the so-called *Lévy stochastic area* process, arising as the area enclosed by the curve of a two-dimensional standard Brownian motion and the chord joining the endpoint to the origin. This process has of course many other interesting properties quite beyond the scope of this work and therefore excluded from it. At this place let us mention the relation of the stochastic area process to the Atiyah-Singer theorems (cf. Bismut (1984, 1988), Léandre (1988), Yor (1991)). There is another (equivalent) definition of $\mathbb{H}$ which is a little different from the one we use. With this other definition, "standard" Brownian motion has an interesting physical interpretation as the joint distribution of a Brownian motion on $\mathbb{R}$, another Brownian motion on $\mathbb{R}$ acting as a random constant field of forces, and the energy produced by the motion (cf. Hulanicki (1976)).

Another interesting subject concerning probabilities on the Heisenberg group (e.g.) is the definition and geometric characterization of an analogue of the Cauchy distribution (cf. Dunau, Sénateur (1986), (1988), Dani (1991)). The work of Neuenschwander (1993) has now been generalized to all positively graduated simply connected nilpotent Lie groups (cf. Neuenschwander, Schott (1996)).

The origin of the Heisenberg group lies in quantum mechanics. It became clear that the

Heisenberg commutation relation reflecting the Heisenberg uncertainty principle can be interpreted within the framework of Lie algebras. Consider the location operator $A : f \mapsto xf$ and the momentum operator $B : f \mapsto \frac{1}{2\pi i} f'$ on the space of infinitely differentiable functions with compact support on $\mathbb{R}$. Then $[A, B] := AB - BA = \frac{1}{2\pi i} I$ (where I is the identity operator). Now the Heisenberg Lie algebra may be interpreted as the algebra generated by A, B, and I. So $\mathbb{H}$ can be described as $\mathbb{R}^3$, equipped with the group multiplication

$$
\begin{aligned}
x \cdot y &= x + y + \frac{1}{2}[x, y] \\
[x, y] &= (0, 0, x'y'' - x''y') \\
&\quad (x = (x', x'', x'''), y = (y', y'', y''')).
\end{aligned}
$$

The center of $\mathbb{H}$ is the line $\{0\} \times \{0\} \times \mathbb{R}$. Compared to the whole class of simply connected nilpotent Lie groups, $\mathbb{H}$ has several special features which are significant for probability theory; we mention:

- $\mathbb{H}$ is stratified,
- if k is the homogeneous dimension and ℓ the class of nilpotency, then $k - 2 - \ell \geq 0$,
- the center is 1-dimensional (so results known for $\mathbb{R}$ may be applied),
- $Aut(\mathbb{H})$ and the (contracting) one-parameter automorphism groups are explicitly known (cf. Folland (1989), pp.19ff., Drisch, Gallardo (1984)) (this can be used in the context of stable and semi-stable semigroups).
- $E([X, c]) = [E(X), c]$, hence $E(X \cdot (-E(X))) = 0$, i.e. $\mathbb{H}$-valued random variables may be centered with their expectation and $\{\prod_{n=1}^{N} X_n\}_{N \geq 1}$, where $\{X_n\}_{n \geq 1}$ are independent and $E(X_n) = 0$, is a martingale,
- $\prod_{j=1}^{n} x_j + \prod_{j=1}^{n} x_{n+1-j} = 2\sum_{j=1}^{n} x_j$ (this has to be used in several places in order to apply results for the vector space-case).
- the density function of the central component of Brownian motion is explicitly known,
- the central component of standard Brownian motion on $\mathbb{H}$ - Lévy's stochastic area process - is a quadratic Brownian functional which has certain relations to winding numbers of two-dimensional Brownian motion (this was used by Shi (1995) to prove his results, which will be presented in 2.3.1).

So the Heisenberg group is much more than just a simple example of a simply connected nilpotent Lie group. $\mathbb{H}$ plays a role in several branches of mathematics and physics, see Folland (1989), Taylor (1986), and the very exhaustive survey article of Howe (1980). One can generalize several notions and facts from harmonic analysis and geometry to $\mathbb{H}$ (see e.g. Korányi (1983, 1985), Taylor (1986)). Relations to combinatorics of paths in the square lattice $\mathbb{Z}^2$ are discussed in Béguin, Valette, and Zuk (1995). There are

also certain applications in signal theory (cf. Schempp (1988)).
Now let us give a brief outline of the contents of this work:
Chapter 1 collects some facts from probability theory on simply connected nilpotent Lie groups G. The notion of a continuous convolution semigroup of probability measures is important. Since G is strongly root-compact (a notion which is due to Böge (1964)) and has no non-trivial compact subgroups, every infinitely divisible probability measure on G can be embedded into a continuous convolution semigroup. This is important for limit theorems and for defining the several types of domains of attraction. These are the contents of sections 1.1 and 1.2. In section 1.3 we present the necessary tools from potential theory as a preparation for the treatment of the Wiener sausage in 2.2.1 and the Lebesgue needle and related questions in 2.2.2.
In chapter 2 we study Brownian motion on $\mathbb{H}$ and its surroundings under the limit theoretic point of view in some detail. Pap (1994) proved that Gauss measures μ on simply connected nilpotent Lie groups determine uniquely the Gauss semigroup in which they may be embedded, but he left as an open problem if μ is also embeddable into a non-Gaussian continuous convolution semigroup. In section 2.1.1 we prove that for $\mathbb{H}$ this is indeed not the case. This may be viewed as a weak form of the Cramér-Lévy theorem telling that (on $\mathbb{R}$) Gaussian distributions have only Gaussian convolution factors. This result will be applied in 3.1.5 to formulate a "transfer principle" between limit theorems on $\mathbb{H}$ and on $(\mathbb{R}^3, +)$. Section 2.1.2 is devoted to a generalization of the Lindeberg theorem due to Pap and the Ljapunov theorem due to Ohring, while in section 2.1.3 we study the domain of normal attraction of Brownian motion on $\mathbb{H}$, work which has been done by Scheffler. An aspect which has to do with robust statistics (in the sense of outlier resistance) is studied in section 2.1.4: We show that after a certain so-called "intermediate" trimming procedure, domains of attraction of other stable semigroups merge into (loosely speaking) domains of attraction of Brownian motion. This explains in certain situations the influence of extremal terms in random products. We will come back to this topic in 3.1.3. Section 2.2.1 is devoted to a limit theorem for the "Wiener sausage" on $\mathbb{H}$ by Chaleyat-Maurel and Le Gall (1989) and its application to the absorbtion of Brownian motion on $\mathbb{H}$ by randomly thrown small sets on $\mathbb{H}$. In section 2.2.2 we present Gallardo's results concerning recurrence and the generalization of the so-called "Lebesgue needle". In section 2.3.1 we study some local and asymptotic results of iterated logarithm type, some of which are due to the author and to Schott. More precisely, we mention the asymptotic (Lévy-Berthuet-Baldi and Chung-Shi) laws of the iterated logarithm, give a new proof of the local (Lévy-Helmes) one, investigate the modulus of continuity, and carry over a qualitative form of the Erdös-Rényi law of large numbers for Brownian motion. Furthermore we present the results of Chaleyat-Maurel and Le Gall (1989) concerning the Hausdorff measure of the range of Brownian motion on $\mathbb{H}$ and the non-existence of multiple points. Section 2.3.2 treats the Crépel-Roynette law of the iterated logarithm for distributions having (roughly speaking) a $(2+\delta)$th absolute moment $(\delta > 0)$, an analogue of the classical Hartman-Wintner law of the iterated logarithm. In the course of the proof Crépel and Roynette gave an estimation of the speed of convergence in the central limit theorem on $\mathbb{H}$, which is of independent interest. In section 2.3.3 we apply the "subsequence-principle", which, in a general form, has been established by Chatterji

and Aldous, to the law of the iterated logarithm of Crépel-Roynette. This principle is a means of transferring limit theorems of i.i.d. random variables to limit theorems for subsequences of dependent random variables. The corresponding theorem for the Marcinkiewicz-Zygmund strong laws of large numbers will be mentioned in 3.2.1.

Further (weak and strong) limit theorems are presented in chapter 3. Hazod (1993) has shown that on arbitrary contractible locally compact groups there exist so-called strictly *universal distributions* (in the sense of Doeblin), i.e. distributions which are partially attracted by every continuous convolution semigroup. Section 3.1.1 is in some sense a complement to that paper for the case where in the normalizing sequence also shifts are allowed: It is shown that a probability measure on $\mathbb{H}$ is universal (in this wider sense with shifts) iff it is universal on the underlying vector space $(\mathbb{R}^3, +)$. In section 3.1.2 we present the characterization of domains of attraction of (non-Gaussian) stable semigroups due to Scheffler. In section 3.1.3 we come back to the topic of 2.1.4: We characterize limits of "lightly" trimmed products in the domain of attraction of certain stable semigroups by means of some "Lévy construction" similar to that for stable laws themselves. In section 3.1.4 we present the results of Tutubalin (1964). Here, the normalization is performed such that the limit measure is a Gaussian distribution on $(\mathbb{R}^3, +)$ instead of $(\mathbb{H}, \cdot)$. Also, the norming maps are not endomorphisms of $(\mathbb{H}, \cdot)$ and the centering is taken with respect to " + " rather than " $\cdot$ ". The topic of section 3.1.5 are triangular systems of probability measures on $\mathbb{H}$ which are not necessarily rowwise identically distributed. We show that limits of commutative infinitesimal triangular systems of probability measures on $\mathbb{H}$ satisfying some "local centering" condition are always infinitely divisible. As in the i.i.d. case for general simply connected nilpotent Lie groups, one can in this situation formulate a "transfer principle", saying that limit theorems for triangular systems on $(\mathbb{R}^3, +)$ have a canonical counterpart on $(\mathbb{H}, \cdot)$ if the measures within each row commute on $\mathbb{H}$. This transfer principle also holds the other way round if the limit measure is known to be Gaussian by the uniqueness property mentioned in 2.1.1. If it would be known that (as in the euclidean case) also on $\mathbb{H}$ any embeddable probability measure μ_1 determines uniquely the continuous convolution semigroup $\{\mu_t\}_{t>0}$ in which it may be embedded, then this transfer principle would also hold the other way round in general. Furthermore, we show that also limits of non-commutative infinitesimal triangular systems of symmetric probability measures on $\mathbb{H}$ are infinitely divisible. While the classical (Kolmogorov) form of the strong law of large numbers is already known in fairly general situations (see e.g. Furstenberg (1963), Tutubalin (1969), Guivarc'h (1976)), we carry over in section 3.2.1 the strong law of large numbers in the form of Marcinkiewicz and Zygmund. We also apply the subsequence principle of 2.3.3 to this situation. In section 3.2.2 the convergence rates are precised; more accurately, the estimations of Baum-Katz and Hsu-Robbins-Erdös are carried over to $\mathbb{H}$. The Baum-Katz theorem is a precision of the Marcinkiewicz-Zygmund law of large numbers, while the Hsu-Robbins-Erdös theorem characterizes complete convergence in the law of large numbers by the finiteness of the second moment. Section 3.2.3 is devoted to the ergodic theorem and related results. Section 3.2.4 presents other (non-classical) versions of laws of the iterated logarithm for stable and semi-stable semigroups which are not yet covered by 2.3.2; these are due to the author and mainly to Scheffler. In section 3.2.5 we carry over the classical "three-series

theorem" due to Kolmogorov to symmetric random variables on $\mathbb{H}$.
Let us close the introduction with some suggestions for further research. In general, the results collected here are in a form which, up to now, has only been proved for $\mathbb{H}$ (or, somewhat more generally, the step 2-case). So it remains open to generalize them to nilpotent Lie groups of higher step.
Another challenging problem whose treatment is just at the beginning now is to try to get rid of the independence or even i.i.d. assumption, which, up to now, has mostly been imposed, and to consider more general processes on groups, also processes with several parameters. Papers which go into this direction are e.g. Watkins (1989), who generalized Donsker's invariance principle for mixing sequences to Lie groups, and the "approximate martingale" approach on abelian groups by Bingham (1993).
The convergence theory for continuous convolution semigroups (which model processes with independent stationary increments) on groups and their generating distributions is now developped quite well. But a vast field still to be examined would be to find corresponding theorems for convolution hemigroups (which model processes with independent, but not necessarily stationary increments) and their generating families. Important first steps in this context were undertaken by Feinsilver (1978), Siebert (1982), Heyer, Pap (1996), and Pap (1996a, 1996b).
Also the generalization to (infinite-dimensional) Hilbert-Lie groups and to p-adic groups has just been begun (cf. Coşkun (1991), Riddhi Shah (1991, 1995), Telöken (1996)).
In theoretical physics, the so-called quantum and braided Heisenberg groups are of growing interest (see e.g. Feinsilver, Schott (1996), section 5.3). There are several approaches to define stochastic processes on these structures (see e.g. Feinsilver, Franz, Schott (1995a), (1995b), Franz, Schott (1996), and the literature cited there). So the question arises if results valid for the ordinary Heisenberg group carry over to this context in some form.
Finally, a subject which, to our knowledge, has not at all been touched so far are random subsets of groups (with the "Minkowski multiplication" $A \cdot B = \{a \cdot b : a \in A, b \in B\}$ as operation). The reason is two-fold: First, limit theorems for random subsets of $\mathbb{R}^d$ with Minkowski addition are often only valid for convex sets, or at least their proof passes by this special case. Now what is a reasonable analogue of convexity on a group? The second reason is that for random sets on $\mathbb{R}^d$, the Minkowski addition of convex subsets corresponds to the addition of their support functions, so one can use the corresponding Banach space results. See e.g. Molchanov (1993), chapter 2 and the literature cited there. This method does not seem to be applicable at all in the non-commutative case.

Chapter 1

Probability theory on simply connected nilpotent Lie groups

1.1 Continuous convolution semigroups of probability measures

Let G be a locally compact group, e the neutral element, $G^* := G\setminus\{e\}$. $(M^1(G), *, \xrightarrow{w})$ is the topological semigroup of (regular) probability measures on G, equipped with the operation of convolution and the weak topology (cf. Heyer (1977), Theorem 1.2.2). A continuous convolution semigroup $\{\mu_t\}_{t>0}$ of probability measures on G (c.c.s.[1] for short) is a continuous semigroup homomorphism

$$([0, \infty[, +) \ni t \mapsto \mu_t \in (M^1(G), *, \xrightarrow{w}),$$

$$\mu_0 = \varepsilon_e$$

(ε_x denotes the Dirac probability measure at $x \in G$.) For simply connected nilpotent Lie groups the request $\mu_0 = \varepsilon_e$ is no restriction, since in any case μ_0 has to be an idempotent element of G and is thus the Haar measure ω_K on some compact subgroup $K \subset G$ (cf. Heyer (1977), 1.5.6); however, simply connected nilpotent Lie groups have no non-trivial compact subgroups (cf. Nobel (1991), 2.2). Let $M^b(G)$ be the Banach algebra of bounded Radon measures on G, equipped with the norm $||.||$ of total variation. For $\mu \in M^b(G)$ one defines

$$\exp \mu := \varepsilon_e + \sum_{k=1}^{\infty} \frac{\mu^{*k}}{k!}.$$

A *Poisson semigroup* is a c.c.s. of the form

$$\{\exp t(\eta - ||\eta||\varepsilon_e)\}_{t>0}.$$

A probability measure on G is called Poisson, if it lies in a Poisson semigroup on G. Let G be a Lie group, $C_b(G)$ the space of all bounded complex-valued functions on

[1]Subsequently, the term "c.c.s." will always mean a continuous convolution semigroup of probability measures.

G, $C_0(G)$ the subspace of all complex-valued continuous functions on G vanishing at infinity, $C_b^\infty(G)$ the space of bounded complex-valued C^∞-functions on G, $\mathcal{D}(G)$ the subspace of complex-valued C^∞-functions with compact support. For a non-negative measure η on G, its adjoint measure $\tilde{\eta}$ is given by $\int_G f(x)\tilde{\eta}(dx) = \int_G f(x^{-1})\eta(dx)$ $(f \in C_b(G))$. The measure η is called symmetric if $\eta = \tilde{\eta}$. For a measurable map $\Phi : G \to G$, the measure $\Phi(\eta)$ is given by $\int_G f(x)\Phi(\eta)(dx) := \int_G F(\Phi(x))\eta(dx)$. A ($G$-valued) random variable X is called symmetric if its law $\mathcal{L}(X) \in M^1(G)$ is symmetric. For $\mu \in M^1(G)$ define the (right) convolution operator $T_\mu : C_b(G) \to C_b(G)$ by

$$T_\mu f(x) := \int_G f(xy)\mu(dy).$$

Now for a c.c.s. $\{\mu_t\}_{t\geq 0}$ and $f \in C_0(G)$ the *infinitesimal generator* $\mathcal{N}$ is defined as

$$\begin{aligned} \mathcal{N}f &:= \lim_{t\to 0+} \frac{1}{t}(T_{\mu_t}f - f) \\ &= \frac{d}{dt}\Big|_{t=0+} T_{\mu_t}f. \end{aligned}$$

$\mathcal{N}$ exists at least on $\mathcal{D}(G)$ (cf. Heyer (1977), Theorem 4.2.8). The whole domain of definition will be denoted by $D_\mathcal{N} \subset C_0(G)$. For $f \in C_b(G)$, the generating distribution $\mathcal{A}$ is defined as

$$\begin{aligned} \mathcal{A}f &:= \lim_{t\to 0+} \frac{1}{t} \int_G [f(x) - f(e)]\mu_t(dx) \\ &= \frac{d}{dt}\Big|_{t=0+} \int_G f(x)\mu_t(dx). \end{aligned}$$

It exists on the whole of $C_b^\infty(G)$ (cf. Siebert (1981), p.119). If $f \in D_\mathcal{N} \cap C_b^\infty(G)$, then $\mathcal{A}f = \mathcal{N}f(0)$.

Now let G be a simply connected nilpotent Lie group. This means that G is a Lie group with Lie algebra $\mathcal{G}$ such that $\exp : \mathcal{G} \to G$ is a diffeomorphism and that the *descending central series* is finite, i.e. there is some $r \in \mathbb{N}_0$ such that

$$\mathcal{G}_0 \supsetneq \mathcal{G}_1 \supsetneq \cdots \supsetneq \mathcal{G}_r = \{0\},$$

where

$$\mathcal{G}_0 := \mathcal{G}, \quad \mathcal{G}_{k+1} := [\mathcal{G}, \mathcal{G}_k] \quad (0 \leq k \leq r-1).$$

G is then called step r-nilpotent. We may (and will from now on often) identify G with $\mathcal{G} = \mathbb{R}^d$ via exp. If it will be necessary to distinguish between objects (elements, functions, ...) on G resp. $\mathcal{G}$, then for an object Ξ on G the corresponding object on $\mathcal{G}$ will be denoted by $^\circ\Xi$. So G may be interpreted as $\mathbb{R}^d$ equipped with a Lie bracket $[.,.] : \mathbb{R}^d \times \mathbb{R}^d \to \mathbb{R}^d$ which is bilinear, skew-symmetric, and satisfies the *Jacobi identity*

$$[[x,y],z] + [[y,z],x] + [[z,x],y] = 0.$$

Write $ad(x)(y) := [x, y]$. The group product is then given by the *Campbell-Hausdorff formula* (cf. Serre (1965)), where due to the nilpotency only the terms up to order r arise:

$$x \cdot y = \sum_{n=1}^{r} z_n,$$

$$z_n = \frac{1}{n} \sum_{p+q=n} (z'_{p,q} + z''_{p,q}),$$

$$z'_{p,q} = \sum_{\substack{p_1+p_2+\ldots+p_m=p \\ q_1+q_2+\ldots+q_{m-1}=q-1 \\ p_i+q_i \geq 1 \\ p_m \geq 1}} \frac{(-1)^{m-1}}{m} \frac{ad(x)^{p_1} ad(y)^{q_1} \ldots ad(x)^{p_m}(y)}{p_1! q_1! \ldots p_m!}.$$

$$z''_{p,q} = \sum_{\substack{p_1+p_2+\ldots+p_{m-1}=p-1 \\ q_1+q_2+\ldots+q_{m-1}=q \\ p_i+q_i \geq 1}} \frac{(-1)^{m+1}}{m} \frac{ad(x)^{p_1} ad(y)^{q_1} \ldots ad(y)^{q_{m-1}}(x)}{p_1! q_1! \ldots q_{m-1}!}.$$

The first few terms are

$$x \cdot y = x + y + \frac{1}{2}[x, y] + \frac{1}{12}\{[[x, y], y] + [[y, x], x]\} + \ldots .$$

It is clear that the neutral element e is 0 and that $x^{-1} = -x$. We will implicitly use the relation $[x + y, y] = [x, y]$. If $x_1, x_2, \ldots \in G$ we will use, for ordered products, the notation $\prod_{j=1}^{n} x_j := x_1 \cdot x_2 \cdot \ldots \cdot x_n$. The simply connected step 2-nilpotent Lie groups play a special role. Here $[.,.]$ is simply a bilinear skew-symmetric map $\mathbb{R}^d \times \mathbb{R}^d \to \mathbb{R}^d$ satisfying $[[x, y], z] = 0$. The most prominent examples are the so-called *Heisenberg groups* $\mathbb{H}^d$, given by $\mathbb{R}^{2d+1}$ and the Lie bracket

$$\begin{aligned} [x, y] \;:=\; & (0, 0, \langle x', y' \rangle - \langle x'', y' \rangle) \\ & (x = (x', x'', x'''), y = (y', y'', y''') \in \mathbb{R}^d \times \mathbb{R}^d \times \mathbb{R} \cong \mathbb{R}^{2d+1}). \end{aligned}$$

This notation (for $d = 1$) will be kept throughout this work. The center of $\mathbb{H}^d$ is the line $\{0\} \times \{0\} \times \mathbb{R}$. The group that we will consider in this work is the three-dimensional Heisenberg group $\mathbb{H}^1$. From now on, we will denote it simply by $\mathbb{H}$. For the central component we will use the notation $q(x) := x'''$ $(x = (x', x'', x''') \in \mathbb{H} \cong \mathbb{R}^3)$. On the other hand we put $p(x) := (x', x'')$. Observe that by the Cauchy-Schwarz inequality we have $|||x, y||| \leq ||x|| \cdot ||y||$ $(x, y \in \mathbb{H})$. This relation will also be used implicitly. $\mathbb{H}$ is the simplest one among the simply connected nilpotent Lie groups, even the simplest non-commutative non-discrete Lie group among all Lie groups. So we think that, generally, in going to the non-commutative situation in Lie group theory, one naturally passes by such groups. Every simply connected step 2-nilpotent Lie group is a quotient of a free simply connected step 2-nilpotent Lie group (cf. Taylor (1986), p.156). We mention that all so-called *groups of type H* (cf. Kaplan (1980)) arising in the context of composition of quadratic forms are simply connected step 2-nilpotent. It turns out that the nilpotent part of the Iwasawa decomposition of a semisimple Lie

group of real rank 1 is of type H (cf. Korányi (1985), Proposition 1.1).
Let G be a simply connected step r-nilpotent Lie group. A positive graduation of $G \simeq \mathcal{G}$ is a vector space decomposition

$$G \sim \mathcal{G} \simeq \bigoplus_{j=1}^{r} V_j$$

such that $[V_i, V_j] \subset V_{i+j}(i+j \leq r)$ and $[V_i, V_j] = \{0\}(i+j > r)$. G is called *stratified* if it admits a positive graduation such that V_1 generates $\mathcal{G}$ as a Lie algebra. The number

$$k := \sum_{i=1}^{r} i \dim(V_i)$$

is called the ***homogeneous dimension*** of G. Clearly, $\mathbb{H}$ is stratified with homogeneous dimension $k = 4$. For $t > 0$, let the dilatations $\delta_t : G \to G$ on a positively graduated simply connected nilpotent Lie group G be given by

$$G \cong \bigoplus_{j=1}^{r} V_j \ni (x_1, x_2, \ldots, x_r) \mapsto (tx_1, t^2x_2, \ldots, t^r x_r) \in \bigoplus_{j=1}^{r} V_j \simeq G$$

(see also Folland, Stein (1982)). Then a *homogeneous norm* on G is a continuous function $|.| : G \to [0, \infty[$ satisfying the properties

(i) $|0| = 0, |x| > 0 (x \in G \backslash \{0\})$.

(ii) $|\delta_t(x)| = t|x| \quad (t > 0, x \in G)$.

It is well-known that all homogeneous norms are equivalent (in the sense that $|.|_1 \leq C|.|_2 \leq D|.|_1$ uniformly on G (cf. Goodman (1977), Lemma 1)). There always exists a homogeneous norm $|.|$ such that

(iii) $|x| = |-x|$ (symmetry),

(iv) $|x \cdot y| \leq |x| + |y|$ (subadditivity)

(cf. Hebisch, Sikora (1990)). For $G = \mathbb{H}$ the homogenous norm

$$|x| := (||(x', x'')||^4 + 16x'''^2)^{1/4} \tag{1.1}$$

is subadditive and symmetric (cf. Korányi (1985), (1.4)). For any homogeneous norm $|.|$ on G there exist constants $C_1, C_2, c > 0$ such that

$$\begin{aligned} |x \cdot y| &\leq C_1(|x| + |y|), \\ |-x| &\leq C_2|x|, \\ ||x_j|| &\leq c|x|^j \end{aligned} \tag{1.2}$$

(cf. Goodman (1977), Lemma 2, Pap (1992), (2)).
Let $p > 0, c \in G$, and X a G-valued random variable. Since

$$E|X \cdot c|^p \leq C_1^p E(|X| + |c|)^p,$$

it follows that

$$E|X|^p < \infty \Longrightarrow E|X \cdot c|^p < \infty.$$

On exponential Lie groups (these are the Lie groups for which $\exp : G \to \mathcal{G} \simeq \mathbb{R}^d$ is a diffeomorphism, so in particular all simply connected nilpotent Lie groups are exponential), the infinitesimal generator and the generating distribution get a very explicit form (if one identifies, as usual, G with $\mathcal{G} \simeq \mathbb{R}^d$ via exp): A distribution $\mathcal{A}$ on $C_b^\infty(G)$ is a generating distribution of a c.c.s. $\{\mu_t\}_{t>0}$ iff it has the form (*Lévy-Hinčin formula*)

$$\mathcal{A}f = \langle \xi, \nabla \rangle f(0) + \langle \nabla, M \cdot \nabla \rangle f(0) + \int_{G^*} [f(x) - f(0) - \Psi(f, x)] \eta(dx), \tag{1.3}$$

where

$$\Psi(f, x) : \begin{cases} \langle x, \nabla \rangle f(0) & : \quad ||x|| \leq 1, \\ \langle \frac{x}{||x||}, \nabla \rangle f(0) & : \quad ||x|| > 1 \end{cases}$$

($f \in C_b^\infty(G)$), $\xi \in G \cong \mathcal{G} \sim \mathbb{R}^d$, M is a positive semidefinite $d \times d$-matrix, and η is a *Lévy measure* on G^*, i.e. a non-negative measure on G^* satisfying

$$\int_{0<||x||\leq 1} ||x||^2 \eta(dx) + \eta(\{x \in G : ||x|| > 1\}) < \infty.$$

ξ, M, η are uniquely determined by $\{\mu_t\}_{t>0}$. (Cf. Siebert (1973), Satz 1). For short, we will write $\mathcal{A} = (\xi, M, \eta)$. $\mathcal{A}$ on $C_b^\infty(G)$ determines uniquely the c.c.s. $\{\mu_t\}_{t\geq 0}$, for this reason we may write $\mu_t =: \operatorname{Exp} t\mathcal{A}$ $(t \geq 0)$; on the other hand, every triple (ξ, M, η) as above generates a c.c.s. (cf. Siebert (1973), Satz 1). $\mathcal{A}$ is called *primitive* if it is of the form $(\xi, 0, 0)$. It is called *Gaussian* if $\eta = 0$. In contrast, it is called *without Gaussian component* if $M = 0$. $\{\operatorname{Exp} t\mathcal{A}\}_{t>0}$ is called Gaussian resp. without Gaussian component if $\mathcal{A}$ has this property. A measure $\mu \in M^1(G)$ is called Gaussian if $\mu = \mu_1$ for some Gaussian c.c.s. $\{\mu_t\}_{t>0}$. A (G-valued) random variable X is called Gaussian if its law $\mathcal{L}(X) \in M^1(G)$ is Gaussian. *Standard Brownian motion* on a simply connected nilpotent Lie group G is the c.c.s. with generating distribution

$$\mathcal{A}f = \Delta_{V_1} f(0)$$

(the so-called *Kohn Laplacian*), where Δ_{V_1} is the Laplace operator with respect to some fixed basis of V_1. Processes corresponding to Gaussian c.c.s. have continuous trajectories. The Poisson semigroup $\{\exp t(\eta - ||\eta||\varepsilon_0)\}_{t>0}$ has generating distribution

$$\mathcal{A}f = \int_{G^*} [f(x) - f(0)] \eta(dx) \quad (f \in C_b^\infty(G))$$

(cf. Siebert (1981), p.119). We can see that the structure of the group does not influence the generating distribution of a c.c.s. This yields the important fact that the map

$$\mathcal{A} \mapsto {}^\circ\mathcal{A},$$

(where ${}^\circ\mathcal{A}({}^\circ f) := \mathcal{A}f$) between generating distributions on G resp. $\mathcal{G}$ is a bijection. This, however, is of course not true for the infinitesimal generator: this one takes the form

$$\mathcal{N}f(x) = \langle \xi, \nabla_G \rangle f(x) + \langle \nabla_G, M \cdot \nabla_G \rangle f(x) + \int_{G^*} [f(xy) - f(x) - \Psi(f, x, y)] \eta(dy) \quad (f \in D_{\mathcal{N}} \cap C_b^{\infty}(G)),$$

where

$$\Psi(f, x, y) := \begin{cases} \langle y, \nabla_G \rangle f(x) & : \quad \|y\| \leq 1, \\ \langle \frac{y}{\|y\|}, \nabla_G \rangle f(x) & : \quad \|y\| > 1, \end{cases}$$

ξ, M, η are the same parameters as in (1.3) and ∇_G is the right gradient $\nabla_G = (E_1, E_2, \ldots, E_d)$ on G, where

$$E_i f(x) := \lim_{t \to 0} \frac{1}{t} [f(x \cdot te_i) - f(x)] \tag{1.4}$$

($e_i = (\delta_{i1}, \delta_{i2}, \ldots, \delta_{id})$, δ_{ij} the Kronecker symbol ($i = 1, 2, \ldots, d$)).
A locally compact group is called *strongly root compact* if for every compact subset $K \subset G$ there exists a compact subset $K_0 \subset G$ such that for every $n \geq 1$ and every sequence $x_1, x_2, \ldots, x_n \in G$ with $x_n = e$ we have

$$Kx_i \cdot Kx_j \cap Kx_{i+j} \neq \emptyset \quad (i + j \leq n) \Longrightarrow x_1, x_2, \ldots, x_n \in K_0$$

(cf. Heyer (1977), Definition 3.1.10; this definition is originally due to Böge (1964)). Strong root compactness has fundamental consequences for probability theory, since in this case for every w-relatively compact subset $N \subset M^1(G)$ also the *root set*

$$R(N) := \bigcup_{\mu \in N} \bigcup_{n \geq 1} \{\nu^m : \nu \in M^1(G), \nu^n = \mu; 1 \leq m \leq n\}$$

is w-relatively compact (cf. Heyer (1977), Theorem 3.1.13). This has the following consequence concerning limit theorems:

Proposition 1.1 *Let G be strongly root compact and assume G has no non-trivial compact subgroups. Suppose $\{\nu_n\}_{n \geq 1} \subset M^1(G)$, $\mu \in M^1(G)$, $\{k(n)\}_{n \geq 1} \subset \mathbb{N}$ is strictly increasing. Then if*

$$\nu_n^{*k(n)} \xrightarrow{w} \mu \quad (n \to \infty),$$

it follows that

$$\nu_n \to \varepsilon_e \quad (n \to \infty)$$

and there exists a c.c.s. $\{\mu_t\}_{t>0}$ such that $\mu = \mu_1$ and a subsequence $\{n(j)\}_{j>1} \subset \mathbb{N}$ such that

$$\nu_{n(j)}^{*\lfloor k(n(j))t \rfloor} \xrightarrow{w} \mu_t \quad (j \to \infty)$$

for every $t \geq 0$.

(Cf. Nobel (1991), Proposition 1, Theorem 1.)
A probability measure $\mu \in M^1(G)$ is called *infinitely divisible* if for any $n \in \mathbb{N}$ there exists $\mu_n \in M^1(G)$ with $\mu_n^{*n} = \mu$. Proposition 1.1 has the following consequence:

Corollary 1.1 *Let μ be an infinitely divisible probability measure on a strongly root compact group G with no non-trivial compact subgroup. Then μ is embeddable into a c.c.s.*

As we already mentioned, simply connected nilpotent Lie groups G have no non-trivial compact subgroups. It follows from Heyer (1977), Theorem 3.1.17 (since $x^n = nx \quad (x \in G, n \in \mathbb{N})$) that they are also strongly root compact. So we have

Corollary 1.2 *Every infinitely divisible probability measure on a simply connected nilpotent Lie group is embeddable into a c.c.s.*

Corollary 1.2 was first proved by Burrell and McCrudden (1974).
A special class of c.c.s. is given by the *stable semigroups* (cf. Hazod (1982, 1984a, 1984b, 1986), Hazod, Siebert (1986), see e.g. Breiman (1968), pp.199ff. for the classical case): Let $\{\tau_t\}_{t>0} \subset Aut(G)$ be a continuous one-parameter group of automorphisms of G (i.e. $t \mapsto \tau_t(x)$ is continuous for every $x \in G$). A c.c.s. $\{\operatorname{Exp} t\mathcal{A}\}_{t>0}$ is called *stable* with respect to $\{\tau_t\}_{t>0}$ if for every $t > 0$ there is a primitive distribution b_t such that

$$\tau_t(\mathcal{A}) = t\mathcal{A} + b_t,$$

where $\tau_t(\mathcal{A})f := \mathcal{A}(f \circ \tau_t)$. It is called *strictly stable* with respect to $\{\tau_t\}_{t>0}$ if $b_t = 0$ for every $t \geq 0$. It is easy to see that strict stability can be characterized directly in terms of the measures themselves: $\{\mu_t\}_{t>0}$ is strictly stable iff

$$\mu_{ts} = \tau_t(\mu_s) \quad (t, s \geq 0).$$

For general stable semigroups this is more complicated: If $\mathcal{A}$ and b_t do not commute (in the sense of the corresponding infinitesimal generators as operators), then $\operatorname{Exp}(t\mathcal{A} + b_t)$ is given by the *Lie-Trotter product formula*:

$$\operatorname{Exp}(t\mathcal{A} + b_t) = w - \lim_{n \to \infty} (\operatorname{Exp} \frac{t}{n}\mathcal{A} \cdot \operatorname{Exp} \frac{1}{n} b_t)^{*n}$$

(cf. Hazod (1977), p.34). Observe that if the μ_t are symmetric, then stability and strict stability coincide. Since in the vector space case $(\mathbb{R}^d, +)$ the distributions $\mathcal{A}$ and b_t commute, it follows at once that the above definition of stability is equivalent to the classical one (operator stability, which was introduced by Sharpe (1969)). If G is exponential, then $\{\tau_t\}_{t>0}$ has to be of the form $\{t^A\}_{t>0}$ for some derivation A of $\mathcal{G}$. In general, A is not uniquely determined even in the Euclidean case (cf. Hudson, Mason (1981)). A generalization of stable semigroups are the so-called *semistable semigroups* (cf. Hazod (1984a)): Let $\tau \in Aut(G)$ and $c > 1$. Then $\{\operatorname{Exp} t\mathcal{A}\}_{t>0}$ is called *semistable* with respect to (τ, c), if there is a primitive distribution b such that

$$\tau(\mathcal{A}) = c\mathcal{A} + b.$$

It is called *strictly semistable* if $b = 0$. Also here, strict semistability may be expressed in terms of the measures themselves: $\{\mu_t\}_{t>0}$ is strictly (τ, c)-semistable if

$$\mu_{ct} = \tau(\mu_t) \quad (t \geq 0).$$

Clearly, every [strictly] stable c.c.s. is a [strictly] semistable c.c.s. A probability measure μ on G is called [strictly] [semi-]stable if it is embeddable into a [strictly] [semi-] stable c.c.s. Domains of attraction of c.c.s. will be examined in the next section. It has been proved by Hazod and Siebert (1986, 1988)) that strictly semistable c.c.s. are always concentrated on a simply connected nilpotent Lie group admitting a positive graduation.

A concept which is more general than that of a c.c.s. (and which will be used in 3.1.5) is the notion of a continuous convolution hemigroup (c.c.h.[2]) of probability measures on a locally compact group G. A c.c.h. $\{\mu_{s,t}\}_{t>s>0} \subset M^1(G)$ is a continuous mapping

$$S := \{(s,t) \in \mathbb{R}^2 : t \geq s \geq 0\} \ni (s,t) \mapsto \mu_{s,t} \in (M^1(G), \xrightarrow{w})$$

satisfying

$$\mu_{r,s} * \mu_{s,t} = \mu_{r,t} \quad (t \geq s \geq r \geq 0),$$

$$\mu_{s,s} = \varepsilon_e \quad (s \geq 0).$$

While c.c.s. represent stochastically continuous processes with independent stationary increments, c.c.h. model non-stationary ones ($\mu_{s,t}$ is the law of the increment $X_s^{-1}X_t$). See the detailed studies of Feinsilver (1978), Siebert (1982), Heyer, Pap (1996), and Pap (1996a, 1996b). This exposition is based on Siebert (1982). Let $G \simeq \mathbb{R}^d$ be a simply connected nilpotent Lie group, $\mu \in M^1(G)$. Then we write

$$q(\mu) := \sum_{k=1}^{d} \Big| \int_{\|x\|\leq 1} x_k \mu(dx) \Big| + \int_G \frac{\|x\|^2}{1+\|x\|^2} \mu(dx) \quad (x = (x_1, x_2, \ldots, x_d) \in \mathbb{R}^d \cong G).$$

A c.c.h. on G is called Lipschitz continuous, if there is a constant $C > 0$ such that

$$q(\mu_{s,t}) \leq C(t-s) \quad (t \geq s \geq 0).$$

Let $\tilde{C}_2(G)$ be the Banach space of all uniformly continuous functions (with respect to the left uniform structure of G) on G vanishing at infinity for which $\tilde{E}_i \tilde{E}_j f$ exists for all $i, j \in \{1, 2, \ldots, d\}$ (where

$$\tilde{E}_i f(x) := \lim_{t \to 0+} \frac{1}{t} [f(te_i \cdot x) - f(x)],$$

$e_i = (\delta_{i1}, \delta_{i2}, \ldots, \delta_{id})$ $(i = 1, 2, \ldots, d)$). The norm $\|.\|_2$ of $\tilde{C}_2(G)$ is given by

$$\|f\|_2 = \|f\| + \sum_{i=1}^{d} \|\tilde{E}_i f\| + \sum_{i=1}^{d} \sum_{j=1}^{d} \|\tilde{E}_i \tilde{E}_j f\|.$$

By formula (1.3), every generating distribution $\mathcal{A}$ of a c.c.s. on G may be extended in a natural way to $\tilde{C}_2(G)$. This extension will also be denoted by $\mathcal{A}$, and the set of

[2] As for c.c.s., the abbreviation "c.c.h." will mean a continuous convolution hemigroup of probability measures.

all such extensions of generating distributions of c.c.s. on G to $\tilde{C}_2(G)$ by $A(G)$. A mapping

$$[0,\infty[\ni t \mapsto \mathcal{A}_t \in A(G)$$

will be called admissible, if for every $f \in \tilde{C}_2(G)$ the map

$$[0,\infty[\ni t \mapsto \mathcal{A}_t f$$

is measurable and bounded. By Siebert (1982), Theorem 4.3, a c.c.h. $\{\mu_{s,t}\}_{t>s>0}$ is Lipschitz continuous iff there exists an admissible mapping $t \mapsto \mathcal{A}_t$ such that the evolution equation

$$\int_G [f(x) - f(0)]\mu_{s,t}(dx) = \int_s^t \mathcal{A}_\sigma(T_{\mu_{\sigma,t}} f)d\sigma \qquad (f \in \tilde{C}_2(G), (s,t) \in S)$$

holds. $\mathcal{A}_t$ is uniquely determined for almost all $t \geq 0$ (cf. Hewitt, Stromberg (1965), Theorem 18.3). $\{\mathcal{A}_t\}_{t>0}$ will be called generating family of $\{\mu_{s,t}\}_{t>s>0}$. On the other hand, for every admissible mapping $t \mapsto \mathcal{A}_t$ there exists exactly one c.c.h. which fulfills the above evolution equation (cf. Siebert (1982), Theorem 5.7). Let $\{\mu_t\}_{t\geq 0} = \{\operatorname{Exp} t\mathcal{A}\}_{t>0}$ be a c.c.s. of symmetric probability measures on G. Clearly, the c.c.h. $\{\mu_{s,t}\}_{t\geq s\geq 0}$ given by

$$\mu_{s,t} = \mu_{t-s}$$

is Lipschitz continuous with $\mathcal{A}_t \equiv \mathcal{A}$ for almost all $t \geq 0$ (differentiate the evolution equation with respect to t at $t = s$, cf. Hewitt, Stromberg (1965), Theorem 18.3).

1.2 Limit theorems

Let G always denote a simply connected nilpotent Lie group. A very general limit theorem was already formulated as Proposition 1.1. In this section we are going to examine:

- Domains of attraction,
- universal distributions,
- the Lindeberg theorem.

The following observation is crucial for weak limit theorems involving c.c.s.:

Proposition 1.2 *Let* $\{\operatorname{Exp} t\mathcal{A}_n\}_{t>0}$ $(n \geq 1), \{\operatorname{Exp} t\mathcal{A}\}_{t\geq 0}$ *be c.c.s. on* G. *Then the following conditions are equivalent:*

(i) $\operatorname{Exp} t\mathcal{A}_n \xrightarrow{w} \operatorname{Exp} t\mathcal{A} \quad (n \to \infty) \quad (t \geq 0)$,

(ii) $\mathcal{A}_n f \to \mathcal{A} f \quad (n \to \infty) \quad (f \in C_b^\infty(G))$.

(Cf. Hazod, Scheffler (1993), Proposition 2.1.a). (ii)$\Longrightarrow$ (i) of Proposition 1.2 is due to Hazod (cf. Hazod (1977), Satz I.2.3), while the converse is implicitly contained in Siebert (1981), which was realized by Hohlov.)

Also of great importance is the following *Poisson approximation theorem* (*accompanying laws theorem*) :

Proposition 1.3 *Let $\{\mu_t\}_{t>0}$ be a c.c.s. on G, $\mu, \nu_n \in M^1(G)(n \geq 1)$, $\{k(n)\}_{n>1} \subset \mathbb{N}$ strictly increasing.. Consider the following conditions:*

(i) $\nu_n^{*[k(n)t]} \xrightarrow{w} \mu_t \quad (n \to \infty) \quad (t \geq 0)$.

(ii) $\exp k(n)t(\nu_n - \varepsilon_0) \xrightarrow{w} \mu_t \quad (n \to \infty) \quad (t \geq 0)$.

(iii) $\nu_n^{*k(n)} \xrightarrow{w} \mu \quad (n \to \infty)$.

(iv) $\exp k(n)(\nu_n - \varepsilon_0) \xrightarrow{w} \mu \quad (n \to \infty)$.

Then we have (i) $\Longleftrightarrow$ (ii) and (iii) $\Longleftrightarrow$ (iv). If G is a euclidean space, then (iii)/(iv) imply the existence of a c.c.s. $\{\mu_t\}_{t>0}$ such that $\mu = \mu_1$ and (i)/(ii) hold.

(Cf. Nobel (1991), Remark 2, our Proposition 1.1, and the fact that for G a euclidean space, the c.c.s. $\{\mu_t\}_{t\geq 0}$ is uniquely determined by μ_1.) Probably (iv) $\Rightarrow$ (ii) also holds in the general case. This can not be proved as long as we do not know the "uniqueness property", i.e. if c.c.s. $\{\mu_t\}_{t>0}$ on simply connected nilpotent Lie groups are uniquely determined by μ_1.

Propositions 1.2 and 1.3 (applied to both $(G, \cdot)$ and $(\mathcal{G}, +)$) together with the identification $\mathcal{A} \leftrightarrow^{\circ} \mathcal{A}$ immediately imply the following "transfer principle", telling that each functional limit theorem for triangular systems of rowwise i.i.d. random variables on $(G, \cdot)$ has a counterpart on $(\mathcal{G}, +)$ and vice versa:

Corollary 1.3 *For $\nu_n \in M^1(G)$, $\{k(n)\}_{n>1} \subset \mathbb{N}$ strictly increasing we have*

$$\nu_n^{*[k(n)t]} \xrightarrow{w} \operatorname{Exp} t\mathcal{A} \quad (t \geq 0)$$

iff

$$({}^{\circ}\nu_n)^{*[k(n)t]} \xrightarrow{w} \operatorname{Exp} t^{\circ}\mathcal{A} \quad (n \to \infty) \quad (t \geq 0).$$

By Proposition 1.3, the "if"-direction also holds if convergence only for one t is stipulated; again, in order to prove the analogue for the "only if"-direction in general, the uniqueness property for c.c.s. on G would be needed.

Definition 1.1 *Let $\{\mu_t\}_{t>0}$ be a c.c.s., $\nu \in M^1(G)$, $N \subset Aut(G)$. Then ν is said to lie in the N-domain of attraction of $\{\mu_t\}_{t>0}$ (symbolically $\nu \in DOA(\{\mu_t\}_{t\geq 0}, N)$) with norming sequence $\{(\tau_n, x_n)\}_{n>1} \subset N \times G$ if*

$$\tau_n(\nu * \varepsilon_{x_n})^{*[nt]} \xrightarrow{w} \mu_t \quad (n \to \infty) \quad (t \geq 0). \tag{1.5}$$

If $\{\mu_t\}_{t\geq 0}$ is $\{t^A\}_{t>0}$-stable, then ν is said to be in the domain of normal attraction with respect to $\{t^A\}_{t>0}$ (symbolically $\nu \in DONA(\{\mu_t\}_{t>0}, \{t^A\}_{t>0})$) if in the norming sequence one has $\tau_n = n^{-A}$. If (1.5) does not hold for the whole sequence $\{n\}_{n\geq 1}$, but only for a subsequence $\{k(n)\}_{n>1}$, then we speak of domains of partial attraction (DOPA resp. DONPA). Strict domains of [partial] attraction (SDO[P]A resp. SDON[P]A) are defined as above with $x_n = 0$ $(n \geq 1)$ in the norming sequence. Domains of attraction of one measure (instead of a c.c.s.) are defined likewise.

Definition 1.2 *Let $A \in Aut(G)$, $\nu \in M^1(G)$. Then ν is called a [strictly] A-universal distribution if $\nu \in [S]DOPA(\{\mu_t\}_{t\geq 0}, \{A^n\}_{n>1})$ for every c.c.s. $\{\mu_t\}_{t>0}$. It is called [strictly] universal if $\nu \in [S]DOPA(\{\mu_t\}_{t\geq 0}, Aut(G))$.*

Universal distributions were (for $G = I\!R$) introduced by Doeblin (1940), A-universal distributions by Nguyen (1981). $A \in Aut(G)$ is called *contracting* if $A^n \to 0 \quad (n \to \infty)$. The following theorem is due to Hazod (Hazod (1993), Theorem 3.2):

Theorem 1.1 *Let $A \in Aut(G)$ be contracting. Then there exists a strictly A-universal probability measure ν.*

For the proof we refer to Hazod (1993). However we indicate how ν is constructed: Let $\{\lambda_m\}_{m \geq 1}$ be a fixed countable dense subset of $\{\lambda \in M^1(G) : \operatorname{supp} \lambda \subset \{x \in G : \|x\| < 1\}\}$. Let $\{\alpha(n)\}_{n \geq 1}, \{\beta(n)\}_{n \geq 1}, \{\gamma(n)\}_{n \geq 1} \subset]0, \infty[$ be sequences such that

(i) $\sum_{n=1}^{\infty} \alpha(n) = 1$,
(ii) $\frac{1}{\alpha(n)} \in I\!N \quad (n \geq 1), \quad \alpha(n) \downarrow 0 \quad (n \to \infty)$,
(iii) $\frac{1}{\alpha(n)} \sum_{k=n+1}^{\infty} \alpha(k) \to 0 \quad (n \to \infty)$,
(iv) $\beta(n) \in I\!N \quad (n \geq 1), \quad \beta(n) \uparrow \infty \quad (n \to \infty)$,
(v) $\gamma(n) := \beta(n) - \beta(n-1) \uparrow \infty \quad (n \to \infty)$,
(vi) $\frac{\rho^{\gamma(n)}}{\alpha(n)} \to 0 \quad (n \to \infty) \quad (\rho \in]0,1[)$:

(one checks that e.g. the classically used sequences

$$\alpha(n) \sim c \cdot 2^{-n^2}, \beta(n) \sim n^3, \gamma(n) \sim 3n^2 - 3n + 1$$

(cf. Hazod (1993)) fulfil (i)-(vi)). Then

$$\nu := \sum_{n=1}^{\infty} \alpha(n) A^{-\beta(n)}(\lambda_n)$$

has the desired properties.
Now we want to formulate an analogue of the Lindeberg theorem on stratified nilpotent Lie groups (cf. Pap (1992), Theorem 3):

Theorem 1.2 *Let $G \cong I\!R^d$ be a stratified step r-nilpotent Lie group with positive graduation $\bigoplus_{j=1}^{r} V_j$ and assume $|.|$ is a homogeneous norm on G. Let $\{\mu_{n,k}\}_{n \geq 1; 1 \leq k \leq k(n)} \subset M^1(G)$ be a triangular system of probability measures on G which is commutative, i.e.*

$$\mu_{n,i} * \mu_{n,j} = \mu_{n,j} * \mu_{n,i} \quad (1 \leq i, j \leq k(n)).$$

Suppose furthermore that the following conditions hold:

(i) $\sup_{n \geq 1} \sum_{k=1}^{k(n)} \int_G |x|^2 \mu_{n,k}(dx) < \infty$,
(ii) $\int_G x_i \mu_{n,k}(dx) = 0 \quad (i = 1, 2, \ldots, p)$,
(iii) $\lim_{n \to \infty} \sum_{k=1}^{k(n)} \int_G x_i \mu_{n,k}(dx) = b_i \quad (i = p+1, p+2, \ldots, q)$,
(iv) $\lim_{n \to \infty} \sum_{k=1}^{k(n)} \int_G x_h x_\ell \mu_{n,k}(dx) = m_{h,\ell} \quad (h, \ell = 1, 2, \ldots, p)$,
(v) $\lim_{n \to \infty} \sum_{k=1}^{k(n)} \int_{|x| \geq \epsilon} |x|^2 \mu_{n,k}(dx) = 0 \quad (\epsilon > 0)$

($x = ((x_1, x_2, \ldots, x_p), (x_{p+1}, x_{p+2}, \ldots, x_q), (x_{q+1}, x_{q+2}, \ldots), (\ldots), \ldots (\ldots x_d)) \in \bigoplus_{j=1}^{r} V_j \cong G \cong I\!R^d$). Then

$$\mu_{n,1} * \mu_{n,2} * \ldots \mu_{n,k(n)} \xrightarrow{w} \operatorname{Exp} tA \quad (n \to \infty),$$

where $\mathcal{A} = (b, M, 0)$,

$$b = (0, 0, \ldots, b_{p+1}, b_{p+2}, \ldots, b_q, 0, 0, \ldots, 0),$$

$$M = \begin{pmatrix} M_p & 0 \\ 0 & 0 \end{pmatrix},$$

$$M_p = (m_{h,\ell})_{1<h,\ell<p}.$$

Proof: Assume $G \cong \mathcal{G} \cong \mathbb{R}^d$. One has to verify the following 4 conditions:

(I) $\lim_{n\to\infty} \sum_{k=1}^{k(n)} \mu_{n,k}(\{x \in G : |x| \geq \varepsilon\}) = 0 \quad (\varepsilon > 0)$,

(II) $\sup_{n\geq 1} \sum_{k=1}^{k(n)} |\int\limits_{|x|<1} x_i \mu_{n,k}(dx)| < \infty \quad (i = 1, 2, \ldots, d)$,

(III) $\lim_{n\to\infty} \sum_{k=1}^{k(n)} \int\limits_{|x|<1} x_i \mu_{n,k}(dx) = \begin{cases} b_i & : \quad i = p+1, p+2, \ldots, q \\ 0 & : \quad \text{else}, \end{cases}$

(IV) $\lim_{n\to\infty} \sum_{k=1}^{k(n)} \int\limits_{|x|<1} x_h x_\ell \mu_{n,k} = \begin{cases} m_{h,\ell} & : \quad h, \ell = 1, 2, \ldots, p \\ 0 & : \quad \text{else} \end{cases}$

(cf. Pap (1992), Grenander (1963), Siebert (1981), Wehn (1962)). (I) follows trivially from (v). Now for $i = 1, 2, \ldots, p$ we have by (1.2) and (ii)

$$\begin{aligned} |\int\limits_{|x|<1} x_i \mu_{n,k}(dx)| &= |\int\limits_{|x|>1} x_i \mu_{n,k}(dx)| \\ &\leq c \int\limits_{|x|>1} |x| \mu_{n,k}(dx) \\ &\leq c \int\limits_{|x|\geq 1} |x|^2 \mu_{n,k}(dx); \end{aligned}$$

hence for $i = 1, 2, \ldots, p$ (II) and (III) follow from (v). For $i = p+1, p+2, \ldots, d$ we obtain again by (1.2) (since $j \geq 2$)

$$\begin{aligned} |\int\limits_{|x|<1} x_i \mu_{n,k}(dx)| &\leq c \int\limits_{|x|<1} |x|^j \mu_{n,k}(dx) \\ &\leq c \int\limits_G |x|^2 \mu_{n,k}(dx), \end{aligned}$$

hence for $i = p+1, p+2, \ldots, d$ (i) implies (II). For $i = p+1, p+2, \ldots, q$ (III) follows from (iii) and (v) using (1.2). For $i = q+1, q+2, \ldots, d$ we have, for any $\varepsilon \in]0, 1[$, by (1.2), (i), and (v) (since $j \geq 3$)

$$\begin{aligned} \limsup_{n\to\infty} \sum_{k=1}^{k(n)} |\int\limits_{|x|<1} x_i \mu_{n,k}(dx)| &\leq \limsup_{n\to\infty} \sum_{k=1}^{k(n)} |\int\limits_{|x|<\varepsilon} x_i \mu_{n,k}(dx)| \\ &\leq c\varepsilon^{j-2} \sup_{n\geq 1} \sum_{k=1}^{k(n)} \int\limits_G |x|^2 \mu_{n,k}(dx) \\ &\to 0 \quad (\varepsilon \to 0). \end{aligned}$$

Hence for $i = q+1, q+2, \dots, d$ (III) follows. Similar estimations (by distinguishing $h, \ell \in \{1, 2, \dots, p\}$ from the other cases) together with (iv) and (v) yield (IV) in all cases. □

The analogue of the classical central limit theorem for stratified groups may be formulated as follows:

Theorem 1.3 *Let $G \cong \mathbb{R}^d$ be a stratified simply connected nilpotent Lie group with positive graduation $\bigoplus_{j=1}^{r} V_j$ and assume $\mu \in M^1(G)$ with*

$$\int_G \overline{x_1}\mu(dx) = 0.$$

$$\int_G \|\overline{x_j}\|^{2/j}\mu(dx) < \infty \quad (j \geq 2)$$

($x = (\overline{x_1}, \overline{x_2}, \dots, \overline{x_r}) \in \bigoplus_{j=1}^{r} V_j$). Then

$$\delta_{n^{-1/2}}(\mu^{*n}) \xrightarrow{w} \operatorname{Exp} \mathcal{A} \quad (n \to \infty),$$

where $\mathcal{A} = (b, M, 0)$,

$$b = (0, \int_G \overline{x_2}\mu(dx), 0, 0, \dots, 0) \in \bigoplus_{j=1}^{r} V_j.$$

$$M = \begin{pmatrix} M_p & 0 \\ 0 & 0 \end{pmatrix},$$

$$M_p = (m_{h,\ell})_{1 \leq h,\ell \leq p},$$

$$m_{h,\ell} = \int_G x_h x_\ell \mu(dx) \quad (h, \ell = 1, 2, \dots, p).$$

$$\overline{x_1} = (x_1, x_2, \dots, x_p) \in V_1.$$

(Cf. Pap (1991b).)

A fact which is also important in connection with limit theorems is the so-called *convergence of types theorem* (cf. Hazod, Nobel (1989), the following considerations are based on). In the classical situation ($G = \mathbb{R}$) it tells the following (see e.g. Loève (1977), p. 216, Breiman (1968), Theorem 8.32):

Theorem 1.4 *Assume $X, X', X_1, X_2, \dots$ are $\mathbb{R}$-valued random variables, X, X' non-degenerate, $a_n > 0, b_n \in \mathbb{R}$ $(n \geq 1)$. If*

$$\mathcal{L}(X_n) \xrightarrow{w} \mathcal{L}(X)$$

and

$$\mathcal{L}(a_n X_n + b_n) \xrightarrow{w} \mathcal{L}(X') \quad (n \to \infty),$$

then

$$\mathcal{L}(X') = \mathcal{L}(aX + b)$$

with $a_n \to a, b_n \to b$ $(n \to \infty)$.

In the non-abelian case, we will not take into consideration the shifts b_n, b. The counterpart of the non-degeneracy on $\mathbb{R}$ is here the notion of *L-S-fullness* ("Linde-Siegel-fullness", cf. Nobel (1988), p. 28): Let G be a simply connected nilpotent Lie group. Then $\mu \in M^1(G)$ is called L-S-full if it is not concentrated on a proper closed connected subgroup of G. Now the convergence of types theorem for automorphisms reads as follows:

Theorem 1.5 *Let G be a simply connected nilpotent Lie group, $\{\tau_n\}_{n>1} \subset Aut(G)$. Assume λ, μ are L-S-full probability measures on G and let $\{\lambda_n\}_{n>1} \subset M^1(G)$. If*

$$\lambda_n \xrightarrow{w} \lambda$$

and

$$\tau_n(\lambda_n) \xrightarrow{w} \mu,$$

then $\{\tau_n\}_{n\geq 1}$ is relatively compact in $Aut(G)$ and for every accumulation point τ of $\{\tau_n\}_{n>1}$ we have $\tau(\lambda) = \mu$.

For the proof of Theorem 1.5 we need some preparations. The first one is a version of the convergence of types theorem on $(\mathbb{R}^d, +)$ (cf. Hazod, Nobel (1989), Lemma 2.1):

Proposition 1.4 *Let $\lambda_n, \lambda, \mu \in M^1(\mathbb{R}^d)$, $\{\tau_n\}_{n>1} \subset Aut(\mathbb{R}^d)$. Assume $\sup_{n\geq 1} ||\tau_n|| = \infty$,*

$$\lambda_n \xrightarrow{w} \lambda,$$

$$\tau_n(\lambda_n) \xrightarrow{w} \mu \quad (n \to \infty).$$

Then λ is not L-S-full.

Proof: W.l.o.g. (by choosing a suitable subsequence) we may assume $||\tau_n|| \to \infty \quad (n \to \infty)$. Put $\sigma_n := ||\tau_n||^{-1}\tau_n$. So again w.l.o.g. we may assume $\sigma_n \to \sigma \in Aut(G) \quad (n \to \infty)$ (with $||\sigma|| = 1$), hence

$$\sigma_n(\lambda_n) \xrightarrow{w} \sigma(\lambda) \quad (n \to \infty).$$

On the other hand, $||\tau_n||^{-1} \to 0$ and

$$\tau_n(\lambda_n) \xrightarrow{w} \mu \quad (n \to \infty)$$

imply

$$\sigma_n(\lambda_n) \xrightarrow{w} \varepsilon_0 \quad (n \to \infty),$$

hence $\sigma(\lambda) = \varepsilon_0$, which means that λ is concentrated on $\ker \sigma \neq \mathbb{R}^d$. □
The following lemma is trivial (cf. Hazod, Nobel (1989), Proposition 1.7):

Lemma 1.1 *For $\mu \in M^1(G)$ we have that μ is L-S-full iff the Poisson measure $\tilde{\mu} := \exp(\mu - \varepsilon_0)$ is L-S-full.*

The next lemma follows from Hazod, Nobel (1989), Theorem 1.11:

Lemma 1.2 *Let $\mu \in M^1(G)$ and assume $\operatorname{supp}\mu$ is a semigroup. Then μ is L-S-full on G iff ${}^{\circ}\mu$ is L-S-full on $(\mathbb{R}^d, +)$.*

Proof: $\Leftarrow$: is trivial.

$\Rightarrow$: Let $\mathcal{V}$ be the subspace of $\mathcal{G}$ generated by $\operatorname{supp}{}^{\circ}\mu \subset \mathcal{G}$. It suffices to show that $\mathcal{V}$ is a subalgebra: Assume ${}^{\circ}x, {}^{\circ}y \in \operatorname{supp}{}^{\circ}\mu$. Since $\operatorname{supp}\mu$ is a semigroup, it follows that ${}^{\circ}(x^k \cdot y^k) \in \operatorname{supp}{}^{\circ}\mu$. But by the Campbell-Hausdorff formula

$$x^k \cdot y^k = kx + ky + \sum_{j=1}^{r} c_j k^{j+1} h_j(x,y)$$

($c_j \in I\!R\setminus\{0\}$, h_j homogeneous polynomials of degree $j+1$). So $k^{-r-1\circ}(x^k \cdot y^k) \in \mathcal{V}$ and thus

$$h_r(x,y) = \lim_{k\to\infty} \frac{1}{c_r} k^{-r-1\circ}(x^k \cdot y^k) \in \mathcal{V}.$$

Hence $\sum_{j=1}^{r-1} c_j h_j(x,y) \in \mathcal{V}$ and by induction we eventually get

$$h_1(x,y) = [x,y] \in \mathcal{V}.$$

So it follows that $[\mathcal{V},\mathcal{V}] \subset \mathcal{V}$. □

Now we are ready to prove Theorem 1.5:

Proof of Theorem 1.5: By Lemma 1.1 we may w.l.o.g. substitute all occurring measures λ_n, etc. by the Poisson measures $\tilde{\lambda}_n$, etc. and thus assume that their supports are semigroups in G. So

$$^{\circ}\lambda_n \xrightarrow{w} {}^{\circ}\lambda,$$

$$^{\circ}\tau_n({}^{\circ}\lambda_n) \xrightarrow{w} {}^{\circ}\mu \quad (n \to \infty).$$

By Lemma 1.2 ${}^{\circ}\lambda$ is L-S-full. Thus by Proposition 1.4 it follows that $\{\tau_n\}_{n\geq 1}$ is relatively compact in $Aut(G)$. Let $\tau \in Aut(G)$ be an accumulation point of $\{\tau_n\}_{n\geq 1}$. Then $\tau(\lambda) = \mu$ follows readily. □

1.3 Some potential theory

Let G always be a stratified nilpotent Lie group of homogeneous dimension $k \geq 3$. In this section we are going to provide the material on which section 2.2.2 will be based. The theory developed in the sequel is due to Gallardo (1982). The aim is to define a sensible notion of capacity for G, to show its geometrical meaning, and to carry over the Wiener test and the Poincaré criterion for immediate entrance of Brownian motion into a set and for recurrence.

Let $\{B(t)\}_{t\geq 0}$ be a realization of standard Brownian motion $\{\mu_t\}_{t\geq 0}$ on G and A some bounded Borel subset of G. Define the first entrance time of $\{B(t)\}_{t\geq 0}$ into A

$$T_A := \inf\{t > 0 : B(t) \in A\}$$

and the regular boundary of A

$$A^r := \{x \in \partial A : P_x(T_A = 0) = 1\},$$

where for $x \in G$

$$P_x(\{B(t)\}_{t\geq 0} \in D) := P(\{xB(t)\}_{t\geq 0} \in D),$$

so e.g. also $P_x(T_A < t) = P(T_{(-x)\cdot A} < t)$, etc. ($E_x(\ldots)$ is defined analogously.) Consider the potential kernel

$$Vf(x) = \int_0^\infty \int_G f(xy)\mu_t(dy)dt$$

and let $\mathcal{B}_c(G)$ be the set of all bounded measurable functions with compact support on G. Let $p_t(x)$ be the density function of μ_t $(t > 0, x \in G)$ (cf. Gallardo (1982), pp. 99f.). Then if one defines

$$|x| := \left(\int_0^\infty p_t(x)dt\right)^{1/(2-k)} \quad (x \in G^*), \tag{1.6}$$

$|.|$ is a symmetric (in the case of $\mathbb{H}$ also subadditive (cf. Gallardo (1982), (3.2))) homogeneous norm and $x \mapsto Vf(x)$ is a bounded continuous function on G such that $Vf(x) \to 0$ $(x \to \infty)(f \in \mathcal{B}_c(G))$. For the proof we refer to Gallardo (1982), Theorem 3.1. The finiteness of $|.|$ is based on a maximum principle due to Bony (1969). By the Hunt theory there exists a finite measure π_A, supported on $A \cup A^r$ such that

$$P_x(T_A < \infty) = V\pi_A(x) = \int_G |(-x)\cdot y)|^{2-k}\pi_A(dy) \tag{1.7}$$

(cf. Gallardo (1982), p.105).

Definition 1.3 *The capacity of A is defined as*

$$C(A) := \pi_A(G).$$

Lemma 1.3 *$C(A)$ is the supremum of all $\mu(G)$, where μ is a finite measure on G with* $\operatorname{supp}\mu \subset A$ *and $V\mu(x) \leq 1$ for all $x \in G$.*

(Cf. Blumenthal, Getoor (1968), p.286, Gallardo (1982), 4.4.)

Proposition 1.5 *It holds that*

$$C(A) = \lim_{x\to\infty} |x|^{k-2}P_x(T_A < \infty).$$

(Cf. Gallardo (1982), Proposition 4.7.)
Proof: The following estimate follows from the definition:

$$\begin{aligned}\inf_{y\in A} |(-x)\cdot y|^{2-k}C(A) &\leq P_x(T_A < \infty)\\ &\leq \sup_{y\in A} |(-x)\cdot y|^{2-k}C(A).\end{aligned}$$

So it suffices to prove

$$\frac{|(-x)\cdot y|}{|x|} \to 1 \quad (x \to \infty)$$

uniformly in $y \in A$. By Gallardo (1982), Lemma 4.7.1 there exist constants $\alpha, C > 0$ such that $||xy| - |x|| \leq \alpha|y|$ for $|x| \geq 2C|y|$. If one writes

$$|\frac{|(-x) \cdot y|}{|x|} - 1| - ||\delta_{1/|x|}(-x) \cdot \delta_{1/|x|}(y)| - |\delta_{1/|x|}(-x)||,$$

the assertion follows. □

Now we show the geometrical meaning of $C(A)$. Denote by λ the Haar measure on G (which is the same as the Lebesgue measure on $I\!R^d$). Consider

$$V_{r,t} := \lambda(\bigcup_{r \leq s \leq t} (A \cdot (-B(s)))),$$

the volume of the region swept out, between times r and t, by the set A attached (in the sense of the group multiplication) to a point obeying to a Brownian motion $\{B(t)\}_{t \geq 0}$. Put $V_t := V_{0,t}$. Then we have:

Theorem 1.6

$$C(A) \overset{a.s.}{=} \lim_{t \to \infty} \frac{V_t}{t}.$$

(Cf. Gallardo (1982), Theorem 4.9).

Proof: We first show that

$$E(V_t) = \int_G P_x(T_A \leq t)dx < \infty.$$

The equality $E(\ldots) = \int \ldots$ follows from the fact that $T_{(-x)\cdot A} \leq t$ iff $x \in \bigcup_{0 \leq s \leq t}(A \cdot (-B(s)))$. Let $\alpha, t > 0$ be fixed and $D := \{x \in G : |x| < r\}$ with $r > 0$ such that $\mu_1(D) = \alpha$. Put $K := \overline{A} \cdot \delta_{t^{1/2}}(D)$. Then we have, for $z \in \overline{A}$ and $0 \leq s \leq t$

$$\begin{aligned}
P_z(B(s) \in K) &= \mu_s((-z) \cdot K) \\
&= \int_K s^{-k/2} \mu_1(\delta_{s^{-1/2}}((-z) \cdot dx)) \\
&= \mu_1(\delta_{s^{-1/2}}((-z) \cdot K))) \\
&\geq \mu_1(\delta_{s^{-1/2}}(\delta_{t^{1/2}}(D))) \\
&= \mu_1(\delta_{(t/s)^{1/2}}(D)) \\
&\geq \mu_1(D) \\
&= \alpha,
\end{aligned}$$

hence

$$\inf_{0 \leq s \leq t} \inf_{z \in \overline{A}} P_z(B(s) \in K) \geq \alpha.$$

Thus we get

$$\begin{aligned}
\mu_t((-x) \cdot K) &= P_x(B(t) \in K) \\
&\geq P_x(B(t) \in K, T_{\overline{A}} \leq t) \\
&= \int_0^t \int_{\overline{A}} P_z(B(t-u) \in K) P_x(B(T_{\overline{A}}) \in dz, T_{\overline{A}} \in du) \\
&\geq \alpha P_x(T_{\overline{A}} \leq t),
\end{aligned}$$

and therefore

$$\begin{aligned}
E(V_t) &= \int_G P_x(T_{\overline{A}} \leq t)dx \\
&\leq \frac{1}{\alpha}\int_G \mu_t((-x)\cdot K)dx \\
&= \frac{\lambda(K)}{\alpha} \\
&< \infty .
\end{aligned}$$

Since the process $\{V_{r,t}\}_{0\leq r\geq t\geq 0}$ is subadditive, it follows by Spitzer (1973), p.905 and the 0-1 law given by Gallardo (1982), Proposition 2.6 (which is based on martingale convergence and a result of Azencott (1970)) that

$$\lim_{t\to\infty}\frac{V_t}{t} \overset{a.s}{=} b.$$

It remains to show that $b = C(A)$. Put

$$e(t) := E(V_t),$$

and, for $\beta > 0$,

$$e^*(\beta) := \int_0^\infty e^{-\beta t} de(t),$$

$$g^\beta(x) := \int_0^\infty e^{-\beta t} p_t(x)dt,$$

$$\Phi^\beta(x) := E_x(e^{-\beta T_A})$$

As above, there exists a finite measure π_A^β supported on $A \cup A^r$ such that

$$\Phi^\beta(x) = \int_G g^\beta((-x)\cdot y)\pi_A^\beta(dy).$$

Let $C^\beta(A) := \pi_A^\beta(G)$. One obtains by an elementary calculation

$$\beta e^*(\beta) = \beta\int_G \Phi^\beta(x)dx = C^\beta(A).$$

By Hunt (1958), p.191 (cf. Gallardo (1982), p.105) $C^\beta(A) \to C(A) \quad (\beta \to 0)$, hence $\beta e^*(\beta) \to C(A) \quad (\beta \to 0)$, which, by the Tauberian theorem, implies $e(t)/t \to C(A) \quad (t \to \infty)$.□

Theorem 2.7 is a related limit theorem concerning the "Wiener sausage" on $\mathbb{H}$. See also Chaleyat-Maurel, Le Gall (1989), pp.257f.

Now we are going to formulate an analogue of Wiener's test (cf. Gallardo (1982), Theorem 5.1). It gives a necessary and sufficient condition for $P_x(T_A = 0)$ being 0 or 1 (observe that by the Blumenthal 0-1 law (see e.g. von Weizsäcker, Winkler (1990),

Corollary 9.7.2) only these two possibilities exist). Let $|.|$ be the homogeneous norm (1.6) and $C \geq 1$ such that

$$|x \cdot y| \leq C(|x| + |y|) \tag{1.8}$$

(as mentioned before, we have $C = 1$ for $G = \mathbb{H}$). Let $\alpha > C$ and assume A is any Borel subset of G. Let x be any fixed point in G and put

$$A_n := \{y \in A : \alpha^{-n-1} < |(-x) \cdot y| \leq \alpha^{-n}\}.$$

Theorem 1.7 *The following conditions are equivalent:*
(i) $P_x(T_A = 0) = 1$,
(ii) $\sum_{n=1}^{\infty} \alpha^{n(k-2)} C(A_n) = \infty$.

Proof: W.l.o.g. we may assume $x = 0$.
(i)$\Longrightarrow$ (ii): Assume that the series converges. Since

$$P_0(T_{A_n} < \infty) \leq \alpha^{(n+1)(k-2)} C(A_n),$$

it follows from the Borel-Cantelli lemma that $T_{A_n} < \infty$ finitely often a.s., hence $P_0(T_A = 0) = 0$.
(ii)$\Longrightarrow$ (i): Suppose (ii) holds. There exists a sequence $\{n(i)\}_{i \geq 1} \subset \{n\}_{n \geq 1}$ with $n(i+1) - n(i) \geq 2$ such that

$$\sum_{i=1}^{\infty} \alpha^{n(i)(k-2)} C(A_{n(i)}) = \infty.$$

Denote by $\theta_s : \Omega \to \Omega$ the shift $B(t, \omega) = B(t - s, \theta_s(\omega))$. Then

$$T_{A_{n(i)}} \circ \theta_{T_{A_{n(j)}}}$$

is the first entrance time in $A_{n(i)}$ after having entered into $A_{n(j)}$. We have

$$\begin{aligned} P_0(T_{A_{n(i)}} < \infty, T_{A_{n(j)}} < \infty) \leq\ & P_0(T_{A_{n(i)}} < \infty, T_{A_{n(j)}} \circ \theta_{T_{A_{n(i)}}} < \infty) \\ & + P_0(T_{A_{n(j)}} < \infty, T_{A_{n(i)}} \circ \theta_{T_{A_{n(j)}}} < \infty). \end{aligned}$$

W.l.o.g. we may assume $j > i$ and we will estimate only the first term on the right hand side of the above inequality, since the second one is analogous. For $x \in \overline{A}_{n(i)}, y \in \overline{A}_{n(j)}$ we have

$$\begin{aligned} |(-x) \cdot y| &= |(-y) \cdot x| \\ &\geq \frac{|x|}{C} - |y| \\ &\geq \frac{\alpha^{-(n(i)+1)}}{C} - \alpha^{-n(j)} \\ &\geq \alpha^{-n(j)}\left(\frac{\alpha}{C} - 1\right). \end{aligned}$$

So

$$
\begin{aligned}
&P_0(T_{A_{n(i)}} < \infty, T_{A_{n(j)}} \circ \theta_{T_{A_{n(i)}}} < \infty) \\
&= \int_{A_{n(i)}} P_x(T_{A_{n(j)}} < \infty) \cdot P_0(B(T_{A_{n(i)}}) \in dx, T_{A_{n(i)}} < \infty) \\
&\leq \sup_{x \in A_{n(i)}} P_x(T_{A_{n(j)}} < \infty) \cdot P_0(T_{A_{n(i)}} < \infty) \\
&\leq \sup_{x \in A_{n(i)}} \int_{\overline{A}_{n(j)}} |(-x) \cdot y|^{2-k} \mu_{A_{n(j)}}(dy) \cdot P_0(T_{A_{n(i)}} < \infty) \\
&\leq (\frac{\alpha}{C} - 1)^{2-k} P_0(T_{A_{n(i)}} < \infty) \cdot \alpha^{n(j)(k-2)} C(A_{n(j)}) \\
&\leq E P_0(T_{A_{n(i)}} < \infty) \cdot P_0(T_{A_{n(j)}} < \infty).
\end{aligned}
$$

Hence we get

$$P_0(T_{A_{n(i)}} < \infty, T_{A_{n(j)}} < \infty) \leq 2E P_0(T_{A_{n(i)}} < \infty) \cdot P_0(T_{A_{n(j)}} < \infty),$$

which allows to apply a result of Lamperti (1963), p.59, with which we get that the probability that infinitely many events

$$T_{A_{n(i)}} < \infty$$

occur is positive, hence 1 by the Blumenthal 0-1 law. By the continuity of the trajectories of $\{B(t)\}_{t>0}$ and its transience (cf. Gallardo (1982), Proposition 3.3 (the proof of which is similar to that of Theorem 1.6 together with an adaption of a method due to Port, Stone (1970), p.162)), (i) follows. □

By a similar method, one can prove a criterion for recurrence (cf. Gallardo (1982), Theorem 7.1). Let A, α be as in Theorem 1.7 and define

$$A_n := \{x \in A : \alpha^n < |x| < \alpha^{n-1}\}.$$

By the 0-1 law Gallardo (1982), 2.6 we have

Lemma 1.4 *Let A be a Borel subset of $G^{[0,\infty[}$ which is invariant with respect to $\{B(t)\}_{t\geq 0}$ (i.e. $\theta_s(A) = A$ for all $s \geq 0$, where θ_s is the shift on the space Σ of trajectories of $\{B(t)\}_{t\geq 0}$: $B(t, \sigma) = B(t - s, \theta_s(\sigma))$). Then*

$$P_0(\textit{for any } t > 0 \textit{ there exists an } s > t \textit{ such that } B(s) \in A) \in \{0, 1\}.$$

Theorem 1.8 *Let A be a Borel subset of G. The following conditions are equivalent:*

(i) $P_0($*for any* $t > 0$ *there exists an* $s > t$ *such that* $B(s) \in A) = 1$,

(ii) $\sum_{n=1}^{\infty} \alpha^{-n(k-2)} C(A_n) = \infty$.

Theorems 1.7 and 1.8 allow to carry over the *Poincaré criterion* to G.

Definition 1.4 *A cone in G with vertex 0 is a Borel subset $C \subset G$ with non-void interior, $0 \in \partial C$, and such that $\delta_r(B) = B$ ($r > 0$). A cone in G with vertex $a \in G$ is a subset of G such that $(-a) \cdot C$ is a cone with vertex 0.*

(Cf. Gallardo (1982), Definition 5.2.) The following Poincaré criterion follows at once from Theorem 1.7 and the fact that $C(\delta_r(A)) = r^{k-2}C(A)$, which is a consequence of Proposition 1.5 (cf. Gallardo (1982), Corollary 5.4):

Corollary 1.4 *Let A be a Borel subset of G, $a \in G$. If there exists a cone C with vertex a and a neighborhood U of a such that $U \cap C \subset A$, then $P_a(T_A = 0) = 1$.*

Similarly one gets by Theorem 1.8 (cf. Gallardo (1982), Corollary 7.3):

Corollary 1.5 *Cones are recurrent sets (in the sense of Theorem 1.8 (i)).*

Chapter 2

Brownian motions on $I\!\!H$

2.1 Weak convergence to Brownian motion

2.1.1 Uniqueness of embedding of Gaussian measures

In the context of c.c.s., also the question if an embeddable probability measure $\mu_1 \in M^1(G)$ determines uniquely the c.c.s. $\{\mu_t\}_{t>0}$ in which it may be embedded is important, since, if μ_1 is embeddable into a unique c.c.s. $\{\mu_t\}_{t\geq 0}$, then, for a strictly increasing sequence $\{k(n)\}_{n\geq 1}$ of natural numbers, the relation $\nu_n^{*k(n)} \xrightarrow{w} \mu_1 \quad (n \to \infty)$ implies $\nu_n^{*[k(n)t]} \xrightarrow{w} \mu_t \quad (t \geq 0) \quad (t \geq 0)$ (cf. Nobel (1991), Remark 2.(a)). It is well-known that this uniqueness property is true for $(I\!\!R^d, +)$. Finite groups satisfy the uniqueness property iff every non-neutral element has order 2 (then the group is of course abelian) (cf. Böge (1959)). For locally compact abelian groups, a sufficient condition for the uniqueness property is the request that the group have no non-trivial compact subgroup (cf. Heyer (1977), Theorem 3.5.15). For irreducible symmetric spaces G/K of noncompact type (i.e. G a semisimple noncompact Lie group with finite center and K a maximal compact subgroup) and K-biinvariant probability measures μ on G Graczyk (1994) used a method to associate to μ a bounded non-negative measure $\tilde{\mu}$ on a Cartan subalgebra $(\mathbf{a}, +)$ such that $\mu_1 * \mu_2 = \mu_3$ iff $\tilde{\mu}_1 * \tilde{\mu}_2 = \tilde{\mu}_3$ and such that $\tilde{\mu}$ determines μ uniquely. This readily yields the uniqueness property for all c.c.s. of K-biinvariant probability measures on G by the uniqueness property on $(\mathbf{a}, +)$. In more general framework, some partial results have been obtained by Hazod (1971). For stable and semistable semigroups on simply connected nilpotent Lie groups see Drisch, Gallardo (1984) and Nobel (1991). A partial result for Poisson semigroups $\{\mu_t^{(i)}\}_{t\geq 0}$ $(i = 1, 2)$ on positively graduated simply connected nilpotent Lie groups has been obtained in Neuenschwander (1995i), Theorem 1: It is shown that if the Lévy measures of both $\{\mu_t^{(i)}\}_{t>0}$ $(i = 1, 2)$ have bounded support, then $\mu_1^{(1)} = \mu_1^{(2)}$ implies $\mu_t^{(1)} = \mu_t^{(2)}$ for all $t \geq 0$. Pap (1994) proved the uniqueness property for the Gauss-semigroups among all Gauss semigroups on simply connected nilpotent Lie groups, generalizing the corresponding result for simply connected step 2-nilpotent Lie groups by Baldi (1985), but he left open the question if Gaussian measures can also be embedded into non-Gaussian c.c.s. The following theorem shows that for Gaussian measures on $I\!\!H$ this is indeed not the case. This may be viewed as a weak form of the Cramér-Lévy theorem telling that (on $\mathbb{R}$) Gaussian distributions have only Gaussian convolution factors. This yields indeed the first non-trivial example of probability measures on non-abelian groups determining uniquely their c.c.s. among all c.c.s. The proof is a combination of the method of Baldi (1985) and a formula of Helmes, Schwane (1983).

Theorem 2.1 *Let $\{\mu_t^{(1)}\}_{t\geq 0}$ be a Gaussian c.c.s. on $\mathbb{H}$ and $\{\mu_t^{(2)}\}_{t\geq 0}$ any c.c.s. on $\mathbb{H}$. Then $\mu_1^{(1)} = \mu_1^{(2)}$ implies $\mu_t^{(1)} = \mu_t^{(2)}$ $(t \geq 0)$.*

For the proof, the following lemma (cf. Baldi (1985), Lemma 4), though elementary, is used crucially:

Lemma 2.1 *Let X be a Gaussian random variable on $\mathbb{R}^n$. Assume A is an $i \times n-$ and B a $j \times n-$matrix. Then there is a $j \times i$-matrix C and a centered Gaussian random variable Z on $\mathbb{R}^j$ which is independent of AX such that $BX = CAX + Z$.*

Proof of Theorem 2.1: Let λ denote Lebesgue measure on $\mathbb{R}^2$. Assume $\mu_t^{(1)} = \mathcal{L}(X_t^{(1)}) = \operatorname{Exp} t(\xi^{(1)}, M^{(1)}, 0), \mu_t^{(2)} = \mathcal{L}(X_t^{(2)}) = \operatorname{Exp} t(\xi^{(2)}, M^{(2)}, \eta^{(2)})$ $(t \geq 0)$. Decompose the matrices

$$M^{(i)} =: \begin{pmatrix} p(M^{(i)}) & c^{(i)} \\ (c^{(i)})^{tr} & q(M^{(i)}) \end{pmatrix} \quad (i = 1,2),$$

where $p(M^{(i)})$ are 2×2-matrices, $c^{(i)} \in \mathbb{R}^2$ $(i = 1,2)$. Put

$$\overline{M^{(i)}} := \begin{pmatrix} p(M^{(i)}) & 0 \\ 0 & 0 \end{pmatrix},$$

$\mathcal{L}(\overline{X_t^{(i)}}) := \operatorname{Exp}(p(\xi^{(i)}), \overline{M^{(i)}}, 0)$ $(i = 1,2)$. By the uniqueness of convolution roots of infinitely divisible probability measures on $(\mathbb{R}^2, +)$ it follows that $p(\xi^{(1)}) = p(\xi^{(2)})$, $p(M^{(1)}) = p(M^{(2)})$, and $p(\eta^{(2)}) = 0$. W.l.o.g. we may assume that $p(X_1^{(i)}) = p(\overline{X_1^{(i)}})$ is genuinely two-dimensional, for otherwise we are in the euclidean situation. It follows that $\eta^{(2)} = 0 \oplus q(\eta^{(2)})$ is concentrated on $\{0\} \times \{0\} \times \mathbb{R} \subset \mathbb{H}$ and

$$\mathcal{L}(q(\overline{X_1^{(1)}})|p(\overline{X_1^{(1)}}) = x) = \mathcal{L}(q(\overline{X_1^{(2)}})|p(\overline{X_1^{(2)}}) = x) \quad (\text{for } \lambda\text{-almost all } x \in \mathbb{R}^2).$$

On the other hand, by Lemma 2.1 it is easy to see (cf. also Baldi (1985), pp.416f.) that there are centered Gaussian measures $\gamma^{(i)}$ $(i = 1,2)$ on $\mathbb{R}$ such that

$$\begin{aligned} \mathcal{L}(q(X_1^{(1)})|p(X_1^{(1)}) = x) &= \mathcal{L}(q(\overline{X_1^{(1)}})|p(\overline{X_1^{(1)}}) = x) \\ &\quad * \varepsilon_{(c^{(1)})^{tr}(x - p(\xi^{(1)}))} * \varepsilon_{q(\xi^{(1)})} * \gamma^{(1)}, \end{aligned} \tag{2.1}$$

$$\begin{aligned} \mathcal{L}(q(X_1^{(2)})|p(X_1^{(2)}) = x) &= \mathcal{L}(q(\overline{X_1^{(2)}})|p(\overline{X_1^{(2)}}) = x) \\ &\quad * \varepsilon_{(c^{(2)})^{tr}(x - p(\xi^{(2)}))} * \varepsilon_{q(\xi^{(2)})} * \gamma^{(2)} \\ &\quad * \operatorname{Exp}(0, 0, q(\eta^{(2)})) \\ &= \mathcal{L}(q(\overline{X_1^{(1)}})|p(\overline{X_1^{(1)}}) = x) \\ &\quad * \varepsilon_{(c^{(2)})^{tr}(x - p(\xi^{(2)}))} * \varepsilon_{q(\xi^{(2)})} * \gamma^{(2)} \\ &\quad * \operatorname{Exp}(0, 0, q(\eta^{(2)})). \end{aligned} \tag{2.2}$$

Since $\mu_1^{(1)} = \mu_1^{(2)}$, taking expectations in (2.1) and (2.2) yields that for some $\zeta^{(i)} \in I\!R$ $(i = 1, 2)$

$$(c^{(1)})^{tr} x + \zeta^{(1)} = (c^{(2)})^{tr} x + \zeta^{(2)}$$

for λ-almost all $x \in I\!R^2$, hence $c^{(1)} = c^{(2)}$. So there exists a Gaussian measure γ on $I\!H$ such that

$$q(\mu_1^{(1)}) = q(\gamma) * \operatorname{Exp}(q(\xi^{(1)}), q(M^{(1)}), 0), \tag{2.3}$$

$$q(\mu_1^{(2)}) = q(\gamma) * \operatorname{Exp}(q(\xi^{(2)}), q(M^{(2)}), q(\eta^{(2)})). \tag{2.4}$$

By the integration by parts formula (7.3.1) in von Weizsäcker, Winkler (1990) (applied to Brownian motion and a deterministic process), formula (14) in Helmes, Schwane (1983) may be invoked. It yields that the Fourier transform (on $I\!R$) of $q(\gamma)$ has only isolated zeroes (since it is a product of a strictly positive function (exp of a real-valued function) and a function which is analytic in a neighborhood of the real axis). Hence by (2.3) and (2.4) it follows that

$$\operatorname{Exp}(q(\xi^{(1)}), q(M^{(1)}), 0) = \operatorname{Exp}(q(\xi^{(2)}), q(M^{(2)}), q(\eta^{(2)}))$$

(since their Fourier transforms (on $I\!R$) have to coincide everywhere (by continuity)), thus again by the uniqueness of roots of infinitely divisible probability measures on $(I\!R, +)$ it follows that $q(\xi^{(1)}) = q(\xi^{(2)}), q(M^{(1)}) = q(M^{(2)})$, and $q(\eta^{(2)}) = 0$, which yields the result. □

It seems that up to now one is still far away from a general solution of the uniqueness problem for c.c.s. on simply connected nilpotent Lie groupps, though it is generally reckognized as one of the most important problems in this theory. To our feeling, all known approaches have their limitations which prevent them from promising a generalization to all c.c.s. on (at least stratified) simply connected nilpotent Lie groups: Hazod (1971) uses the spectral decomposition (which can only work for normal measures), Drisch, Gallardo's (1984) and Nobel's (1991) results are based on (semi-)stability, whereas Pap (1994) and Neuenschwander (1995i) make heavy use of moments. The proof of Theorem 2.1 does not seem to carry over to generalize either.

2.1.2 The Lindeberg and Ljapunov theorems

Let $\{B_i(t)\}_{t>0} (i = 1, 2)$ be two independent standard Brownian motions on $I\!R$. Then the standard Brownian motion on $I\!H$ may be expressed as:

$$\{B(t)\}_{t\geq 0} = \{(B_1(t), B_2(t), A(t))\}_{t\geq 0},$$

$$A(t) = \frac{1}{2}\int_0^t (B_2 dB_1 - B_1 dB_2)$$

(see e.g. Roynette (1975), Feinsilver, Schott (1989)). $\{B(t)\}_{t\geq 0}$ is a strong Markov process. $\{A(t)\}_{t\geq 0}$ is the so-called *Lévy stochastic area process*. It describes the area enclosed by the curve $\{(B_1(t), B_2(t))\}_{t\geq 0}$ and the chord joining it to the origin. It is well known that $A(1)$ has the Fourier transform (on $I\!R$)

$$\varphi(u) = (\cosh \frac{u}{2})^{-1} \quad (u \in I\!R)$$

and thus the density

$$f(x) = (\cosh \pi x)^{-1} \quad (x \in I\!R).$$

The calculation of the Fourier transform can be done directly by approximating the stochastic integral by Riemann sums (cf. Berthuet (1979)). See also Berthuet (1986), Helmes (1986), Helmes, Schwane (1983) for related results.
The classical Bernstein theorem says that if X, Y are i.i.d. real-valued random variables, then $X + Y$ and $X - Y$ are independent iff $\mathcal{L}(X) = \mathcal{L}(Y)$ is Gaussian. One may ask if such a theorem is also valid on groups. For a discussion of this problem on abelian groups, see Heyer (1977), 5.3 and Feldman (1987). On simply connected nilpotent Lie groups G, $(X \cdot Y, Y \cdot X)$ and $(X \cdot (-Y), (-Y) \cdot X)$ are independent iff $\mathcal{L}(X) = \mathcal{L}(Y)$ is a Gaussian measure concentrated on an abelian subgroup of G (cf. Neuenschwander, Schott (1996)). It can be shown that even for standard Brownian motion on $I\!H$, the random variables $X \cdot Y$ and $X \cdot (-Y)$ are not independent: Let $\{B(t)\}_{t \geq 0}, \{B'(t)\}_{t \geq 0}$ be two independent standard Brownian motions on $I\!H$, $X := B(1)$, $Y := B'(1)$, $T = (T_1, T_2, T_3) := X \cdot Y$, $S = (S_1, S_2, S_3) := X \cdot (-Y)$. Then a direct calculation shows that $E(T_1 T_2 S_3) = -1$, hence (since $E(S_3) = 0$) $T_1 T_2$ and S_3 are negatively correlated, thus T and S are not independent (cf. Neuenschwander, Schott (1996)).
A representation of $A(t)$ in terms of one-dimensional Gaussian distributions can be found in Chaleyat-Maurel (1981), p. 205: Let $A_n, B_n, \tilde{A}_n, \tilde{B}_n$ $(n \geq 1)$ be independent $I\!R$-valued standard Gaussian random variables. Then

$$A(2\pi) = \sum_{n=1}^{\infty} \frac{1}{n}(\tilde{B}_n(A_n - \frac{B_1(2\pi)}{\sqrt{\pi}}) - B_n(\tilde{A}_n - \frac{B_2(2\pi)}{\sqrt{\pi}}).$$

The proof follows from the Fourier development of (ordinary) Brownian motion:

$$B_1(t) = \sum_{n \in \mathbb{Z} \setminus \{0\}} c_n \frac{i}{n} f_n(t) - (\sum_{n \in \mathbb{Z} \setminus \{0\}} c_{-n} \frac{i}{n}) f_0(t) + \frac{c_0 t}{\sqrt{2\pi}},$$

$$c_n = \int_0^{2\pi} \overline{f}_n(s) dB_1(s),$$

$$f_n(t) = \frac{1}{\sqrt{2\pi}} e^{int} \quad (n \in \mathbb{Z})$$

(cf. Chaleyat-Maurel (1981), p.204, Ikeda, Watanabe (1989), p. 476).
Another interesting representation of $A(t)$ (which will be used in 2.3.1) is given in Ikeda, Watanabe (1989), formula (VI.6.6):

Theorem 2.2

$$A(t) = B_3(\frac{1}{4} \int_0^t (B_1(s)^2 + B_2(s)^2) ds),$$

where $\{B_3(t)\}_{t \geq 0}$ is another standard Brownian motion on $I\!R$ independent of $\{\{B_1(t)\}_{t \geq 0}, \{B_2(t)\}_{t \geq 0}\}$.

Proof (cf. Ikeda, Watanabe (1989), pp. 470ff.): Put

$$R(t) := \sqrt{B_1(t)^2 + B_2(t)^2} \tag{2.5}$$

By Itô's formula, it follows that

$$\frac{R(t)^2}{2} = \int_0^t R dB_4 + t, \tag{2.6}$$

where (by Lévy's characterization) $\{B_4(t)\}_{t>0}$ is the ($\mathbb{R}$-valued standard) Brownian motion

$$B_4(t) = \int_0^t (\frac{B_1}{R} dB_1 + \frac{B_2}{R} dB_2).$$

We have $\langle B_4, B_4 \rangle_t = t$, $\langle B_4, A \rangle_t = 0$, and $\langle A, A \rangle_t = \frac{1}{4} \int_0^t R(s)^2 ds$. Define $B_3(t) := A(\Phi_t)$, where Φ_t denotes the inverse function of

$$t \mapsto \frac{1}{4} \int_0^t R(s)^2 ds.$$

By a theorem due to Knight (cf. Ikeda, Watanabe (1989), Theorem II.7.3) $\{B_3(t)\}_{t \geq 0}$ and $\{B_4(t)\}_{t \geq 0}$ are independent. $\{R(t)\}_{t \geq 0}$ is the pathwise unique solution of (2.6) with initial condition $R(0) = 0$. So the σ-field generated by $\{R(s)\}_{0 \leq s \leq t}$ is contained in the σ-field generated by $\{B_4(t)\}_{0 \leq s < t}$, hence $\{B_3(t)\}_{t \geq 0}$ and $\{R(t)\}_{t > 0}$ are independent. □
For another definition of "standard Brownian motion", Hulanicki (1976) gave a physical interpretation (cf. the Introduction).
In this section, a special form of the Lindeberg theorem due to Pap (cf. Pap (1992), Theorem 5) is formulated. For a version vaild on all stratified groups see Theorem 1.2. Let $|.|$ be any homogeneous norm on $\mathbb{H}$. $\{\mu_t\}_{t>0} \subset M^1(\mathbb{H})$ denotes standard Brownian motion on $\mathbb{H}$.

Lemma 2.2 *Let A be a linear endomorphism of $\mathbb{R}^2$. Then*

$$\overline{A} := \begin{pmatrix} A & 0 \\ 0 & \det A \end{pmatrix}$$

is a (group) endomorphism of $\mathbb{H}$.

Proof: It suffices to prove $q(\overline{A}[x, y]) = q([\overline{A}x, \overline{A}y])$. But

$$\begin{aligned} q([\overline{A}x, \overline{A}y]) &= \det(A(x', x''), A(y', y'')) \\ &= \det(A \cdot \begin{pmatrix} x' & y' \\ x'' & y'' \end{pmatrix}) \\ &= \det A \cdot \det \begin{pmatrix} x' & y' \\ x'' & y'' \end{pmatrix} \\ &= \det A \cdot q([x, y]) \\ &= q(\overline{A}[x, y]). \end{aligned}$$

Let A be a positive definite $n \times n$-matrix. Then there exists exactly one positive definite $n \times n$-matrix $\sqrt{A}$ such that $\sqrt{A} \cdot \sqrt{A} = A$ (let D be a diagonal matrix such that $A = SDS^{-1}$ and define $\sqrt{D}$ in the obvious way, then $\sqrt{A} := S\sqrt{D}S^{-1}$).

Theorem 2.3 *Let* $\{\nu_n\}_{n>1} \subset M^1(\mathbb{H})$ *and*

$$\int_{\mathbb{H}} |x|^2 \nu_n(dx) < \infty.$$

Put

$$a_n := \int_{\mathbb{H}} x\nu_n(dx) \in \mathbb{H}.$$

$$A_n := \sum_{k=1}^{n} \int_{\mathbb{H}} \begin{pmatrix} x' \\ x'' \end{pmatrix} \otimes \begin{pmatrix} x' \\ x'' \end{pmatrix} (\nu_k * \varepsilon_{-a_k})(dx).$$

Assume furthermore that

$$(\nu_n * \varepsilon_{-a_n}) * (\nu_m * \varepsilon_{-a_m}) = (\nu_m * \varepsilon_{-a_m}) * (\nu_n * \varepsilon_{-a_n}) \quad (n, m \geq 1).$$

Suppose A_n is positive definite $(n \geq 1)$.

(a) $\sup_{n\geq 1} \frac{1}{\sqrt{\det A_n}} \sum_{k=1}^{n} \int_{\mathbb{H}} |x'''|(\nu_k * \varepsilon_{-a_k})(dx) < \infty.$

(b) $\operatorname{tr}(A_n^{-1}) \sum_{k=1}^{n} \int_{|x|^2 \geq \varepsilon/\operatorname{tr}(A_n^{-1})} |x|^2 (\nu_k * \varepsilon_{-a_k})(dx) \to 0 \quad (n \to \infty) \quad (\varepsilon > 0).$

Then

$$\sqrt{A_n}^{-1}(\nu_1 * \varepsilon_{-a_1} * \nu_2 * \varepsilon_{-a_2} * \ldots * \nu_n * \varepsilon_{-a_n}) \xrightarrow{w} \mu_1 \quad (n \to \infty).$$

Proof: First, observe that for a $\mathbb{H}$-valued random variable X and $c \in \mathbb{H}$ we have $E(X \cdot c) = E(X) \cdot c$, so

$$\begin{aligned} \int_{\mathbb{H}} x(\nu_n * \varepsilon_{-a_n})(dx) &= \int_{\mathbb{H}} x\nu_n(dx) \cdot (-a_n) \\ &= a_n \cdot (-a_n) \\ &= 0. \end{aligned}$$

We only have to verify that the triangular system

$$\{\nu_{n,k}\}_{n\geq 1; 1\leq k\leq n} = \{\sqrt{A_n}^{-1}(\nu_k * \varepsilon_{-a_k})\}_{n\geq 1; 1\leq k\leq n}$$

satisfies conditions (i), (iv), and (v) of Theorem 1.2. Observe that

$$\begin{aligned} ||A_n(x', x'')||^2 &= \langle A_n(x', x''), A_n(x', x'')\rangle \\ &= \langle A_n^2(x', x''), (x', x'')\rangle \\ &\leq ||(x', x'')||^2 \operatorname{tr}(A_n^2) \end{aligned} \tag{2.7}$$

and

$$\det A_n \leq \frac{1}{2}\operatorname{tr}(A_n^2) \tag{2.8}$$

Furthermore, by the equivalence of homogeneous norms

$$|(x', x'', x''')|^2 \leq C(x'^2 + x''^2 + |x'''|) \tag{2.9}$$

for some $C > 0$. Hence by (2.7)-(2.9)

$$\begin{aligned} |\overline{A_n}(x', x'', x''')|^2 &\leq C(||A_n(x', x'')||^2 + \det A_n \cdot |x'''|) \\ &\leq C''|x|^2 \mathrm{tr}(A_n^2). \end{aligned}$$

Thus

$$\begin{aligned} \int_{|x| \geq \epsilon} |x|^2 \nu_{n,k}(dx) &= \int_{|\overline{\sqrt{A_n}^{-1}} x| \geq \epsilon} |\overline{\sqrt{A_n}^{-1}} x|^2 (\nu_k * \varepsilon_{-a_k})(dx) \\ &\leq C' \mathrm{tr}(A_n^{-1}) \int_{|x|^2 \geq \epsilon^2/\mathrm{tr}(A_n^{-1})} |x|^2 (\nu_k * \varepsilon_{-a_k})(dx). \end{aligned}$$

So (b) yields (v) of Theorem 1.2. Since

$$\sum_{k=1}^{n} \int_{\mathbb{H}} \begin{pmatrix} x' \\ x'' \end{pmatrix} \otimes \begin{pmatrix} x' \\ x'' \end{pmatrix} \nu_{n,k}(dx) = (\delta_{ij})_{1 \leq i,j \leq 2}$$

(δ_{ij} the Kronecker symbol), (iv) is also proved. This, together with (a), implies (i). □
We mention that in Pap (1993), section 8, more theorems of this type are available for $\mathbb{H}$. But due to Siebert (as Pap and Hazod mentioned in a personal communication to the author) they are valid in a more general context.
Related to Theorem 2.3 is the following generalization of Ljapunov's theorem due to Ohring (1993).
Let $M^1(\mathbb{H})^\natural$ denote the abelian subsemigroup of $M^1(\mathbb{H})$ consisting of those measures which are invariant with respect to orthogonal transformations of $\mathbb{R} \times \mathbb{R} \times \{0\} \subset \mathbb{H}$ (cf. Ohring (1993), p.1315).

Theorem 2.4 *Let $\{\nu_n\}_{n \geq 1} \subset M^1(\mathbb{H})^\natural$ and let $\{\mu_t\}_{t \geq 0} \subset M^1(\mathbb{H})^\natural$ be standard Brownian motion on $\mathbb{H}$. Assume*

$$\int_{\mathbb{H}} x''' \nu_n(dx) = 0 \quad (n \geq 1),$$

$$\gamma_n := \int_{\mathbb{H}} |x|^3 \nu_n(dx) < \infty \quad (n \geq 1),$$

$$\Gamma_n / s_n^3 \to 0 \quad (n \to \infty),$$

where

$$\Gamma_n := \sum_{j=1}^{n} \gamma_j,$$

$$s_n := (\sum_{j=1}^{n} \sigma_j^2)^{1/2},$$

$$\sigma_n := \left(\int_{\mathbb{H}} (x'^2 + x''^2)\nu_n(dx)\right)^{1/2} \quad (n \geq 1).$$

Put

$$\beta_a(x) := ax \quad (x \in \mathbb{H}, a > 0).$$

Then we have

$$\beta_{s_n^{-1}}(\nu_1 * \nu_2 * \ldots * \nu_n) \xrightarrow{w} \mu_1 \quad (n \to \infty).$$

Proof: Let $X_1, Y_1, X_2, Y_2, \ldots$ be independent $\mathbb{H}$-valued random variables such that $\mathcal{L}(X_n) = \nu_n$ and $\mathcal{L}(Y_n) = \mu_{\sigma_n}$. Put

$$Z_{n,j} := Y_1 \cdot Y_2 \cdot \ldots \cdot Y_{j-1} \cdot X_{j+1} \cdot X_{j+2} \cdot \ldots \cdot X_n \quad (1 \leq j \leq n).$$

Let $f : \mathbb{H} \to \mathbb{H}$ be any fixed C^3-function with bounded derivatives of orders up to 3. Since $M^1(\mathbb{H})^\natural$ is abelian, we get by the Taylor formula on $\mathbb{H}$ (with remainder term of order 3) after some straightforward calculations using the assumptions of the theorem

$$\begin{aligned}
&|E(f(\beta_{s_n^{-1}}(\prod_{j=1}^n X_j))) - E(f(\beta_{s_n^{-1}}(\prod_{j=1}^n Y_j)))| \\
&= |\sum_{j=1}^n |E(f(\beta_{s_n^{-1}}(X_j \cdot Z_{n,j}))) - E(f(\beta_{s_n^{-1}}(Y_j \cdot Z_{n,j})))|| \\
&\leq M {s_n}^{-3} \sum_{j=1}^n (\gamma_j + E|Y_j|^3) \\
&\leq M' {s_n}^{-3} \Gamma_n,
\end{aligned}$$

where the last inequality follows from the fact that by the scaling property of $\{\mu_t\}_{t>0}$ and Hölder's inequality with $p = \frac{3}{2}$ we have

$$E|Y_j|^3 = c\sigma_j^3 \leq c\gamma_j.$$

Hence we have shown that

$$|E(f(\beta_{s_n^{-1}}(\prod_{j=1}^n X_j))) - E(f(Z_n))| = O(\Gamma_n / s_n^3) \quad (n \to \infty),$$

where Z_n are $\mathbb{H}$-valued random variables with $\mathcal{L}(Z_n) = \mu_1$, which proves the assertion.
□

Remark 2.1 *Theorem 2.4 does not follow from Theorem 1.2 since it may happen that*

$$\limsup_{n \to \infty} \sum_{j=1}^n \int_{\mathbb{H}} \frac{|x'''|}{s_n} \nu_n(dx) = \infty,$$

which contradicts condition (i) of Theorem 1.2.

A somewhat related (functional) limit theorem for products on $\mathbb{H}$ of martingale differences converging to Brownian motion on $\mathbb{H}$ is implied by Janson, Wichura (1983), Theorem 1.1 (and (i) on p.76; see also pp.77f.):

Theorem 2.5 *Let, for each $n \geq 1$, $\{S_{n,k}\}_{1<k\leq k(n)}$ be an $\mathbb{R}^2$-valued mean 0 square integrable martingale adapted to increasing σ-fields $\{\mathcal{F}_{n,k}\}_{0\leq k<k(n)}$ and assume $\{\sigma_{n,k}\}_{0\leq k\leq k(n)}$ are $\mathbb{R}$-valued random variables with*

$$0 = \sigma_{n,0} < \sigma_{n,1} < \ldots < \sigma_{n,k(n)} = 1$$

and such that $\sigma_{n,k}$ is $\mathcal{F}_{n,k-1}$-measurable $(1 \leq k \leq k(n))$. Define

$$W(t) = (W_1(t), W_2(t)) := S_{n,\max\{k:\sigma_{n,k}\leq t\}} \quad (0 \leq t \leq 1).$$

Suppose

$$W_n \to (B_1, B_2) \quad (n \to \infty)$$

in the Skorohod topology on the space $\mathcal{D}([0,1],\mathbb{R})$, where (B_1, B_2) is a standard $\mathbb{R}^2$-valued Brownian motion. Put

$$X_{n,k} := S_{n,k} - S_{n,k-1} =: (X_{n,k}^{(1)}, X_{n,k}^{(2)}).$$

Assume further that either

$$\limsup_{n\to\infty} P(\sum_{k=1}^{k(n)} E(X_{n,k}^2 \mid \mathcal{F}_{n,k-1}) \geq c) \to 0 \quad (c \to \infty)$$

or

$$\sum_{\sigma_{n,k}\leq t} E(X_{n,k}^2 \mid \mathcal{F}_{n,k-1}) \xrightarrow{P} t.$$

Then

$$q(\prod_{k=1}^{k(n)} \begin{pmatrix} X_{n,k}^{(1)} \\ X_{n,k}^{(2)} \\ 0 \end{pmatrix}) \xrightarrow{w} \frac{1}{2}\int_0^1 (W_1 dW_2 - W_2 dW_1) = A(1),$$

where $\{A(t)\}_{t\geq 0}$ is Lévy's area process.

2.1.3 The domain of normal attraction

Our purpose in this section is to present a theorem offered by Scheffler (cf. Scheffler (1993), Theorem 5.9) on the strict domain of normal attraction of a strictly stable Gaussian semigroup on $\mathbb{H}$.

Let $|.|$ be any homogeneous norm on $\mathbb{H}$.

Interestingly enough, a difference to the euclidean case arises here: While in the euclidean case a measure lying in the domain of normal attraction of a Gaussian measure necessarily has a finite second moment, this need not be the case on $\mathbb{H}$ (cf. Scheffler (1993), Example 5.10): The measure with density

$$f(x', x'', x''') = \frac{e}{4\pi} e^{-\frac{1}{2}(x'^2 + x''^2)} \frac{\log|x'''| + 1}{x'''^2 \log^2 |x'''|} 1\{|x'''| \geq e\}$$

lies in $SDONA(\{\operatorname{Exp} t\mathcal{A}\}_{t\geq 0}, \{\delta_{1/\sqrt{t}}\}_{t>0})$, where

$$\mathcal{A}f = \frac{\partial^2}{\partial x'^2} f(0) + \frac{\partial^2}{\partial x''^2} f(0)$$

is the Kohn Laplacian, but $\int_{\mathbb{H}} |x|^2 f(x)dx = \infty$. Define the pseudo inner product on $\mathbb{H}$ by

$$h(x,y) := x'y' + x''y'' + \sqrt{|x'''y'''|}.$$

Consider the symplectic form on $\mathbb{R}^2$ given by $\sigma(x,y) = q([(x,0),(y,0)])$. A linear endomorphism $M : \mathbb{R}^2 \to \mathbb{R}^2$ is called a [skew-]symplectic mapping if $\sigma(Mx, My) = [-]\sigma(x,y)$. For $M \in sp(\mathbb{R}^2)$ — Lie algebra of the symplectic group and $m > 0$ define

$$\sigma_{M,m}(t) := \begin{pmatrix} t^{M \cdot m} & 0 \\ 0 & t^{2m} \end{pmatrix}.$$

A measure $\mu \in M^1(\mathbb{H})$ is called *full*, if μ is not concentrated on a conjugacy class of a proper closed connected subgroup of $\mathbb{H}$ (cf. Scheffler (1993), Definition 3.1, Drisch, Gallardo (1984), Proposition 3.8). If for a c.c.s. $\{\mu_t\}_{t>0}$ on $\mathbb{H}$ the measure μ_1 is full, then every μ_t is full (cf. Drisch, Gallardo (1984), p.65). By Drisch, Gallardo (1984), 2.5 and the proof of Corollary 2.6, every contracting one-parameter group $\{\tau_t\}_{t>0} \subset Aut(\mathbb{H})$ (i.e. $\tau_t \to 0$ $(t \to 0)$) is, up to conjugation with an inner automorphism of $\mathbb{H}$, of the form $\{\sigma_{M,m}(t)\}_{t>0}$ $(M \in sp(\mathbb{R}^2), m > 0)$. Moreover, if $\{\mu_t\}_{t\geq 0}$ is full and $\{\sigma_{M,m}\}_{t>0}$-stable, then $m \geq \frac{1}{2}$ (cf. Drisch, Gallardo (1984), Theorem 1). Let $\mathcal{B} \subset Aut(\mathbb{H})$ denote the subgroup of type

$$\begin{pmatrix} mF & 0 \\ 0 & [-]m^2 \end{pmatrix}$$

(F [skew-]symplectic, $m > 0$) (note that every automorphism on $\mathbb{H}$ is of this form up to an inner automorphism (cf. Scheffler (1993), (5), Folland (1989), Theorem 1.22)). For the following theorem see Scheffler (1993), Theorem 5.9:

Theorem 2.6 *Assume $\{\mu_t\}_{t\geq 0}$ is an L-S-full strictly $\{\sigma_{M,m}(t)\}_{t>0}$-stable Gaussian semigroup on $\mathbb{H}$ and let C be the covariance matrix of $p(\mu_1)$. For $\nu \in M^1(\mathbb{H})$ consider the conditions*

(i) $\nu \in SDONA(\{\mu_t\}_{t>0}, \{\sigma_{M,m}(t)\}_{t>0})$,

(ii) $h(((y',y'') \cdot C^{tr}, 0), (y',y'',0)) = \int_{\mathbb{H}} h^2(x,(y',y'',0))\nu(dx) \quad ((y',y'') \in \mathbb{R}^2)$.

Then (i)$\Longrightarrow$ (ii). If ν and μ_1 are symmetric and $\int_{\mathbb{H}} |x|^2 \nu(dx) < \infty$, then also (ii)$\Longrightarrow$ (i).

Proof: (i)$\Longrightarrow$(ii): This is just a consequence of the corresponding statement on $(\mathbb{R}^2, +)$ (cf. Jurek (1980), Theorem 4.1).
(ii)$\Longrightarrow$(i): By the central limit theorem on $\mathbb{H}$ (cf. Theorem 1.3) it follows that

$$\delta_{1/\sqrt{n}}(\nu^{*n}) \xrightarrow{w} \mu_1 \quad (n \to \infty). \tag{2.10}$$

Clearly, $\{\mu_t\}_{t>0}$ is also strictly $\{\delta_{1/\sqrt{t}}\}_{t>0}$-stable. Since

$$\mu_1 \in SDONA(\mu_1, \{\delta_{1/\sqrt{t}}\}_{t>0})$$

and

$$\mu_1 \in SDONA(\mu_1, \{\sigma_{M,m}(t)\}_{t>0}),$$

it follows by the convergence of types theorem (Theorem 1.5) that $\{\sigma_{M,m}(\frac{1}{n})\delta_{\sqrt{n}}\}_{n\geq 1}$ is relatively compact in $\mathcal{B}$ and all accumulation points τ lie in the invariance group of μ_1 (i.e. $\tau(\mu_1) = \mu_1$). Now (2.10) and

$$\sigma_{M,m}(\frac{1}{n})(\nu^{*n}) = ((\sigma_{M,m}(\frac{1}{n})\delta_{\sqrt{n}})\delta_{1/\sqrt{n}})(\nu^{*n})$$

complete the proof. □

2.1.4 A Gaussian central limit theorem for intermediately trimmed products

As an aspect which has to do with robust statistics (in the sense of outlier resistance), trimmed sums on $\mathbb{R}$ and on more general Banach spaces have been given considerable attention in the literature. For a survey on the development of the subject up to about the beginning of 1988 see the Introduction of Häusler (1988). Roughly speaking, a trimmed sum is a sum of n i.i.d. random variables where a certain number $k(n)$ of the most extreme observations has been removed. For $n \to \infty$ one speaks of "light" resp. "heavy" trimming if $k(n)$ is constant resp. proportional to n. One has also looked at the case where $k(n) \to \infty$, but $k(n)/n \to 0 \quad (n \to \infty)$: this is called "intermediate" trimming. For sums of random variables on the real line there is also an other aspect under which trimming procedures may be distinguished: either one prescribes how many of the largest and how many of the smallest summands have to be deleted or, what is then called "modulus trimming", a certain amount of the observations greatest in absolute value are removed. Somewhat surprisingly, for symmetric distributions, the behavior in both cases may be totally different, as is pointed out in Häusler (1988), p.9: For example, there are symmetric laws μ for which the intermediately modulus-trimmed sums of n i.i.d. random variables obeying to the law μ, after suitable normalization, asymptotically have a Gaussian distribution, whereas by deleting the $k(n)/2$ largest and the $k(n)/2$ smallest values ($k(n) \to \infty, k(n)/n \to 0 \quad (n \to \infty)$) one gets a sequence which cannot be normalized to converge to a Gaussian limit. It is clear that results concerning modulus trimming come above all into consideration when they have to be carried over to more general structures such as vector spaces and groups. (See Neuenschwander (1991), Introduction.)
Kuelbs and Ledoux (1986) proved the following assertion for intermediately trimmed sums:
Let $X_1, X_2, \ldots$ be i.i.d. random variables on some type 2-Banach space $\mathbb{B}$ such that

$$S_n = a_n \sum_{k=1}^{n} (X_k + x_n)$$

(with $a_n > 0$ and $x_n \in \mathbb{B}$ suitably chosen) converges weakly to some non-degenerate dilatation-stable random variable on $\mathbb{B}$. Let $\{r(n)\}_{n\geq 1}$ be a sequence such that $r(n) \to \infty, r(n)/n \to 0 \ (n \to \infty)$. For $\xi, \tau > 0$ define ${}^{(\xi,r(n))}S_{n,\tau}$ as the trimmed sum which

arises from S_n by the following procedure: Put $F_n(j) := \sharp\{i : \|X_i\| > \|X_j\|$ for $1 \leq i \leq n$ or $\|X_i\| = \|X_j\|$ for $1 \leq i \leq j\}$ and set $X_n^{(k)} := X_j$ if $F_n(j) = k$ (so, roughly speaking, $X_n^{(k)}$ is the k-th largest element of $\{X_1, X_2, \ldots, X_n\}$); now delete $X_n^{(1)}, X_n^{(2)}, \ldots, X_n^{(\lfloor \xi r(n) \rfloor)}$ provided their norm is larger than $\tau^{1/\alpha}/a_{n/r(n)}$; furthermore replace a_n by $a_{\lfloor n/r(n) \rfloor}/\sqrt{r(n)}$ and x_n by $\tilde{x}_{n/r(n)} := -E(X_1 \cdot 1\{\|X_1\| \leq \tau^{1/\alpha}/a_{\lfloor n/r(n) \rfloor}\})$. Then for every $\tau > 0$ ${}^{(\xi, r(n))}S_{n,\tau}$ tends weakly to some non-degenerate centered Gaussian random variable if one takes ξ big enough.(Cf. Kuelbs, Ledoux (1986), Theorem 2.)
Let $\{t^A\}_{t>0}$ be a contracting one-parameter automorphism group on $\mathbb{H}$. Then there is a "natural" polar coordinate decomposition of $\mathbb{H} = \mathbb{R}^3$, i.e. there is a norm p_A on $\mathbb{R}^3$ such that the mapping

$$]0, \infty[\times S \ni (t, y) \mapsto t^A y = x \in \mathbb{R}^3 \backslash \{0\} = \mathbb{H} \backslash \{0\},$$

where S denotes the unit sphere with respect to p_A, is a homeomorphism (cf. Jurek (1984), Corollary 2). Now for $\mathbb{H}$-valued random variables in the $\{t^A\}_{t>0}$-domain of attraction of a $\{t^A\}_{t>0}$-stable semigroup, where $\{t^A\}_{t>0}$ acts contracting, the trimming and the truncation in the definition of $\tilde{x}_{\lfloor n/r(n) \rfloor}$ will be performed with respect to the "radial component" t rather than the euclidean norm $\|.\|$; we will show that in this situation the above-mentioned assertion due to Kuelbs and Ledoux carries fully over to $\mathbb{H}$ (cf. Neuenschwander (1995a)).
Assume $\mathcal{T} = \{t^A\}_{t>0}$ is a one-parameter automorphism group on $\mathbb{H}$ which acts contracting on $\mathbb{H}$. Let $\mathcal{S} = \{\operatorname{Exp} tA\}_{t>0}$ be a non-degenerate $\mathcal{T}$-stable semigroup on $\mathbb{H}$ with generating distribution (1.3) and suppose $X_1, X_2, \ldots$ are i.i.d. $\mathbb{H}$-valued random variables such that $\nu = \mathcal{L}(X_1) \in DOA(\mathcal{S}, \mathcal{T})$ with norming sequence $\{(t_n^A, x_n)\}_{n>1}$. Let S be the unit sphere in $\mathbb{R}^3$ with respect to the norm p_A mentioned above and consider, for $n \in \mathbb{N}$, the polar coordinate decomposition $X_n = T_n^A Y_n$ ($T_n \geq 0, Y_n \in S$). For $1 \leq j \leq n$ put $F_n(j) := \sharp\{i : T_i > T_j$ for $1 \leq i \leq n$ or $T_i = T_j$ for $1 \leq i \leq j\}$ and set $X_n^{(k)} := X_j$ if $F_n(j) = k$. Let $\{r(n)\}_{n \geq 1}$ be a sequence such that $r(n) \to \infty, r(n)/n \to 0$ $(n \to \infty)$. For $\xi, \tau > 0$ let ${}^{(\xi, r(n))}J_{n,\tau}$ be the set of those indexes $j \in \{1, 2, \ldots, n\}$ for which $X_j = X_n^{(k)}$ with $1 \leq k \leq \lfloor \xi r(n) \rfloor$ and $T_j > \tau/t_{\lfloor n/r(n) \rfloor}$. Put $\tilde{x}_{\lfloor n/r(n) \rfloor} := -E(X_1 \cdot 1\{T_1 \leq \tau/t_{\lfloor n/r(n) \rfloor}\})$. Define the trimmed product

$${}^{(\xi, r(n))}P_{n,\tau} = r(n)^{-1/2} t^A_{\lfloor n/r(n) \rfloor} \Big(\prod_{j=1}^{n} (X_j \cdot 1\{j \notin {}^{(\xi, r(n))}J_{n,\tau}\}) \cdot \tilde{x}_{\lfloor n/r(n) \rfloor}\Big).$$

Theorem 2.7 *Under the above-mentioned conditions, for every* $\tau > 0$, ${}^{(\xi, r(n))}P_{n,\tau}$ *converges weakly to some non-degenerate centered Gaussian random variable on* $\mathbb{H}$ *if one takes* ξ *big enough.*

The proof of Theorem 2.7 is similar to that of the corresponding part of Theorem 2 in Kuelbs, Ledoux (1986).
The following lemma is the analogue of Lemma 3.3 in Kuelbs, Ledoux (1986). Put

$$\nu_n = t_n^A(\nu * \varepsilon_{x_n}),$$

$${}^{r(n)}U_{n,\tau} = r(n)^{-1/2} t^A_{\lfloor n/r(n) \rfloor} \Big(\prod_{j=1}^{n} (X_j \cdot 1\{T_j \leq \tau/t_{\lfloor n/r(n) \rfloor}\}) \cdot \tilde{x}_{\lfloor n/r(n) \rfloor}\Big).$$

Lemma 2.3 *$P(^{(\xi,r(n))}P_{n,\tau} \neq^{r(n)} U_{n,\tau}) \to 0$ $(n \to \infty)$ if ξ is chosen large enough.*

Proof: By Proposition 1.3 we have

$$\exp(nt(\nu_n - \varepsilon_0)) \xrightarrow{w} \operatorname{Exp} t\mathcal{A} \quad (n \to \infty)$$

for all $t \geq 0$. By Siebert (1981), Propositions 6.1, 6.4 there is a basis $\mathcal{U}$ of closed neighborhoods of 0 such that for all $U \in \mathcal{U}$

$$n\nu_n|_{cpl(U)} \xrightarrow{w} \eta|_{cpl(U)} \quad (n \to \infty). \tag{2.11}$$

By the contraction property it is easy to see that (2.11) is possible only if

$$t_n^A(x_n) \to 0. \tag{2.12}$$

By (2.11) and (2.12) we have

$$nt_n^A(\nu)|_{\mathrm{cpl}\,U} \xrightarrow{w} \eta|_{\mathrm{cpl}\,U} \quad (n \to \infty)$$

for all $U \in \mathcal{U}$, hence

$$c := \sup_{n \geq 1} nP(t_n T_1 > \tau) < \infty$$

holds. The rest of the proof proceeds exactly as in the proof of Lemma 3.3 in Kuelbs, Ledoux (1986). Put

$$p_n := P(T_1 > \tau / t_{\lfloor n/r(n) \rfloor}).$$

Then we get by Feller (1968), p.173, and Stirling's formula

$$\begin{aligned}
P(^{(\xi,r(n))}P_{n,\tau} \neq^{r(n)} U_{n,\tau}) &\leq P(\sharp\{j \in \{1,2,\dots,n\} : T_j > \tau/t_{\lfloor n/r(n) \rfloor}\} \geq \lfloor \xi r(n) \rfloor + 1) \\
&= \sum_{j=\lfloor \xi r(n) \rfloor + 1}^{n} \binom{n}{j} p_n^j (1-p_n)^{n-j} \\
&= n \binom{n-1}{\lfloor \xi r(n) \rfloor} \int_0^{p_n} t^{\lfloor \xi r(n) \rfloor} (1-t)^{n - \lfloor \xi r(n) \rfloor - 1} dt \\
&\leq n \cdot n^{\lfloor \xi r(n) \rfloor} p_n^{\lfloor \xi r(n) \rfloor} \int_0^{p_n} (1-t)^{n - \lfloor \xi r(n) \rfloor - 1} dt / \lfloor \xi r(n) \rfloor ! \\
&\leq \left(\frac{np_n e}{\lfloor \xi r(n) \rfloor}\right)^{\lfloor \xi r(n) \rfloor} \lfloor \xi r(n) \rfloor^{-1/2} n/(n - \lfloor \xi r(n) \rfloor) \\
&\leq 2 \cdot \lfloor \xi r(n) \rfloor^{-1/2} \cdot (2cr(n)e/\lfloor \xi r(n) \rfloor)^{\lfloor \xi r(n) \rfloor}
\end{aligned} \tag{2.13}$$

for n large enough. Now (2.13) implies the assertion.□

The following lemma is also of independent interest:

Lemma 2.4 *If $\nu \in DOA(\{\operatorname{Exp} t\mathcal{A}\}, T)$ with norming sequence $\{(t_n^A, x_n)\}_{n \geq 1}$ and T acts contracting on $\mathbb{H}$, then $^\circ\nu \in DOA(\{\operatorname{Exp} t^\circ\mathcal{A}\}_{t \geq 0}, {}^\circ T)$ with norming sequence $\{({}^\circ t_n^A, {}^\circ x_n)\}_{n \geq 1}$.*

Proof: Consider the triangular system $\{Z_{nj}\}_{n\geq 1, 1\leq j\leq 2n}$ of random variables on $I\!H$ given by

$$Z_{nj} := \begin{cases} t_n^A(x_n) & : \ j \text{ even}, \\ t_n^A(X_{(j+1)/2}) & : \ j \text{ odd}. \end{cases}$$

Then

$$\mathcal{L}(\prod_{j=1}^{2n} Z_{nj}) \xrightarrow{w} \operatorname{Exp}\mathcal{A} \quad (n \to \infty). \tag{2.14}$$

By (2.12) (which follows from the contraction property) it follows that

$$\mathcal{L}(\prod_{j=1}^{2n} Z_{n,2n+1-j}) \xrightarrow{w} \operatorname{Exp}\mathcal{A} \quad (n \to \infty). \tag{2.15}$$

too. Now

$$\begin{aligned}
\prod_{j=1}^{2n} Z_{nj} + \prod_{j=1}^{2n} Z_{n,2n+1-j} &= \sum_{j=1}^{2n} Z_{nj} + \frac{1}{2}\sum_{1\leq i<j\leq 2n}[Z_{ni}, Z_{nj}] + \sum_{j=1}^{2n} Z_{n,2n+1-j} \\
&\quad + \frac{1}{2}\sum_{1\leq i<j\leq 2n}[Z_{n,2n+1-i}, Z_{n,2n+1-j}] \\
&= 2\sum_{j=1}^{2n} Z_{nj} + \frac{1}{2}\sum_{1\leq i<j\leq 2n}([Z_{ni}, Z_{nj}] + [Z_{nj}, Z_{ni}]) \\
&= 2\sum_{j=1}^{2n} Z_{nj}.
\end{aligned} \tag{2.16}$$

It follows from (2.14)-(2.16) that the sequence

$$\{\sum_{j=1}^{2n}{}^{\circ} Z_{nj}\}_{n\geq 1}$$

is uniformly tight and thus weakly relatively compact by Prohorov's theorem. By Propositions 1.2 and 1.3 it follows that the sequence $\{{}^{\circ}\mathcal{A}_n\}_{n\geq 1}$ of generating distributions

$${}^{\circ}\mathcal{A}_n({}^{\circ}f) = n\int_G [{}^{\circ}f({}^{\circ}x) - {}^{\circ}f(0)]{}^{\circ}t_n^A({}^{\circ}\nu * \varepsilon_{{}^{\circ}x_n})(d^{\circ}x)$$

is relatively compact in the topology of "pointwise" convergence on $C_b^{\infty}(I\!R^3)$. Let N_{nt} be an $I\!N^{\circ}$-valued random variable which obeys to a Poisson law with parameter nt and which is independent of $\{X_n\}_{n\geq 1}$. By (2.12) it follows that

$$\begin{aligned}
t_n^A(\sum_{j=1}^{N_{nt}}(X_j \cdot x_n)) - t_n^A(\sum_{j=1}^{N_{nt}}(X_j + x_n)) &= \frac{1}{2}[t_n^A(\sum_{j=1}^{N_{nt}}(X_j + x_n)), t_n^A(x_n)] \\
&\xrightarrow{w} \varepsilon_0 \quad (n \to \infty).
\end{aligned} \tag{2.17}$$

Thus by Proposition 1.2 we have that if ${}^{\circ}\mathcal{A}_{n'}({}^{\circ}f) \to {}^{\circ}\tilde{\mathcal{A}}({}^{\circ}f) \quad (n' \to \infty) \quad ({}^{\circ}f \in C_b^{\infty}(I\!R^3))$ for a subsequence $\{n'\} \subset \{n\}$, it follows that

$$\operatorname{Exp} t^{\circ}\tilde{\mathcal{A}}_{n'} \to \operatorname{Exp} t^{\circ}\tilde{\mathcal{A}} \quad (n' \to \infty) \quad (t \geq 0). \tag{2.18}$$

where

$$\check{A}_n(f) := n \int_G |f(x) - f(0)| \nu_n(dx).$$

Relation (2.18) implies, by Proposition 1.2 and the identification $\mathcal{A} \leftrightarrow^{\circ} \mathcal{A}$,

$$\check{A}_{n'}(f) \to \check{A}(f) \quad (n' \to \infty) \quad (f \in C_b^{\infty}(\mathbb{H})).$$

So by Proposition 1.2

$$\operatorname{Exp} t\check{A}_{n'} \xrightarrow{w} \operatorname{Exp} t\check{A} \quad (n' \to \infty) \quad (t \geq 0).$$

But

$$\operatorname{Exp} t\check{A}_n \xrightarrow{w} \operatorname{Exp} t\mathcal{A} \quad (n \to \infty) \quad (t \geq 0)$$

by our assumption and Proposition 1.3, so we must have $\check{A} = \mathcal{A}$. Thus ${}^{\circ}\mathcal{A}_n({}^{\circ}f) \to {}^{\circ}\mathcal{A}({}^{\circ}f)$ $(n \to \infty)$ $({}^{\circ}f \in C_b^{\infty}(\mathbb{R}^3))$, which, together with Propositions 1.2 and 1.3, implies the assertion.□

Now we are ready to prove our theorem.

Proof of Theorem 2.7: By Lemma 2.3 we have to prove that ${}^{r(n)}U_{n,\tau}$ converges weakly to a non-degenerate centered Gaussian limit. By Propositions 1.2, 1.3, the identification $\mathcal{A} \leftrightarrow^{\circ} \mathcal{A}$, and an argument similar to (2.17), it suffices to prove the corresponding assertion on $(\mathbb{R}^3, +)$, i.e. we have to show that

$${}^{r(n)}V_{n,\tau} = r(n)^{-1/2} t^A_{\lfloor n/r(n) \rfloor} \Big(\sum_{j=1}^{n} (X_j \cdot 1\{T_j \leq \tau / t_{\lfloor n/r(n) \rfloor}\}) + \tilde{x}_{\lfloor n/r(n) \rfloor} \Big)$$

tends weakly to a non-degenerate centered Gaussian law on $(\mathbb{R}^3, +)$ as $n \to \infty$. Hence we still have to apply the "if"-part of Araujo,Giné (1980), Theorem 3.5.9 to the triangular system

$$\{ r(n)^{-1/2} t^A_{\lfloor n/r(n) \rfloor} (X_j \cdot 1\{T_j \leq \tau / t_{\lfloor n/r(n) \rfloor}\}) \}_{n \geq 1, 1 \leq j \leq n}$$

on $(\mathbb{R}^3, +)$. For every $\delta > 0$

$$n \cdot P(\| r(n)^{-1/2} t^A_{\lfloor n/r(n) \rfloor} (X_j \cdot 1\{T_j \leq \tau / t_{\lfloor n/r(n) \rfloor}\}) \| \geq \delta) = 0$$

for n large enough, which proves condition (i) of Araujo, Giné (1980), Theorem 3.5.9. In order to prove (ii) of Araujo, Giné (1980), Theorem 3.5.9 we have to verify that

$$\left\{ \begin{matrix} \limsup \\ \liminf \end{matrix} \right\}_{n \to \infty} n \cdot \operatorname{Var}\langle s, r(n)^{-1/2} t^A_{\lfloor n/r(n) \rfloor} (X_1 \cdot 1\{T_1 \leq \tau / t_{\lfloor n/r(n) \rfloor}\}) \cdot$$
$$1\{\| r(n)^{-1/2} t^A_{\lfloor n/r(n) \rfloor} (X_1 \cdot 1\{T_1 \leq \tau / t_{\lfloor n/r(n) \rfloor}\}) \| \leq \delta\} \rangle \to \tilde{Q}(s) \quad (\delta \downarrow 0) \tag{2.19}$$

for some non-degenerate quadratic form $\tilde{Q}(s)$ on $\mathbb{R}^3$. If one puts $m := \lfloor n/r(n) \rfloor$, then (2.19) may be rewritten as

$$\lim_{m \to \infty} m \cdot \operatorname{Var}\langle s, t^A_m(X_1) \cdot 1\{T_1 \leq \tau / t_m\} \rangle = \tilde{Q}(s) \not\equiv 0 \tag{2.20}$$

(observe that in (2.19) for every $\delta > 0$ we have that $1\{\ldots\} = 1$ for n big enough). Now (2.20) follows immediately from Lemma 2.4, Araujo, Giné (1980), Theorems 3.5.9, 3.5.10 (applied to $\mathbb{R}$), and the property

$$\eta(dx) = \lambda\kappa(dy)\frac{dt}{t^2} \qquad (\lambda \geq 0, \kappa \in M^1(S)) \tag{2.21}$$

with $\tilde{Q}(s) = E(\langle s, X\rangle^2)$ (where

$$E(e^{i\langle s,X\rangle}) = \exp[-Q(\langle s, .\rangle) + \int_{[0,\tau]\times S} (e^{i\langle s,t^A y\rangle} - 1 - i\langle s, t^A y\rangle)\eta(dt, dy)])$$

due to the fact that Q and η in (1.3) resp. (2.21) are not both $\equiv 0$ (since $\{\operatorname{Exp} t\mathcal{A}\}_{t>0}$ is supposed to be non-degenerate).□

Remark 2.2 *The fact that ${}^{\tau(n)}U_{n,\tau}$ converges to some non-degenerate centered Gaussian random variable on $\mathbb{H}$ is a limit theorem for truncated $\mathbb{H}$-valued random variables, which is of its own interest.*

2.2 Capacities

2.2.1 The Wiener sausage

The following theorem is Theorem 7.2 in Chaleyat-Maurel, Le Gall (1989) (the whole section will be based on this article). It is a limit theorem for the "Wiener sausage" on $\mathbb{H}$ arising as the set swept out by a fixed set $\epsilon K \subset \mathbb{H} \cong \mathbb{R}^3$ attached (in the euclidean sense) to a point obeying to the shifted standard Brownian motion $\{x \cdot B(t)\}_{t>0}$ on $\mathbb{H}$. Let K be a compact subset of $\mathbb{H} \sim \mathbb{R}^3$ and define, for $\epsilon > 0, 0 \leq s \leq t$

$$S_{\epsilon K}(s, t) := \bigcup_{s\leq u\leq t} (x \cdot B(u) + \epsilon K).$$

Let E', E'', E''' be the left invariant (with respect to $\mathbb{H}$) vector fields (1.4) corresponding to the standard basis vectors $e' = (1, 0, 0), e'' = (0, 1, 0), e''' = (0, 0, 1) \in \mathbb{R}^3 \simeq \mathbb{H}$. For $x \in \mathbb{H}$ denote the projection

$$p_x(y'E'(x) + y''E''(x) + y'''E'''(x)) := y'''$$

and consider the height of K at x given by

$$h_x(K) = \sup_{y\in K} p_x(y) - \inf_{y\in K} p_x(y). \tag{2.22}$$

Furthermore let

$$\begin{aligned} V_x(K) := 2\pi\lambda(\{&y'E'(x) + y''E''(x) + y'''E'''(x) : \\ &0 \leq y' \leq 1, 0 \leq y'' \leq 1, 0 \leq y''' \leq h_x(K)\}) \end{aligned} \tag{2.23}$$

(where λ denotes Lebesgue measure on $\mathbb{R}^3$ (which is the same as Haar measure on $\mathbb{H}$)). Assume R is a bounded domain in $\mathbb{H} \sim \mathbb{R}^3$ and let $\varphi : R \to \mathbb{R}$ be a bounded measurable function. Define $\kappa(dx) := \varphi(x)dx$. Let

$$\tau := \inf\{t \geq 0 : B(t) \notin R\}.$$

Theorem 2.8 *For every $p \geq 1, T > 0, x_0 \in R$ we have*

$$E_{x_0}\Big(\sup_{0<s<t\leq T} \Big| \frac{\log(1/\varepsilon)}{\varepsilon} \kappa(S_{\varepsilon K}(s,t)) - \int_s^{t\wedge T} \varphi(x_0 \cdot B(u)) V_{x_0 \cdot B(u)}(K)\, du \Big|^p \Big) \to 0 \quad (\varepsilon \to 0).$$

For the proof of Theorem 2.8 we need several preparations. The first one is a theorem on capacities (cf. Chaleyat-Maurel, Le Gall (1989), Theorem 6.1). For notation, see section 1.3 (unless otherwise stated).

Theorem 2.9

$$\sup_{x \in R} \Big| \frac{\log(1/\varepsilon)}{\varepsilon} C(x + \varepsilon K) - V_x(K) \Big| \to 0 \quad (\varepsilon \to 0).$$

Let ξ be an exponental random variable of parameter β, independent of $\{B(t)\}_{t>0}$. Define the process $\{B^\beta(t)\}_{t\geq 0}$ by

$$B^\beta(t) := \begin{cases} B(t) & : \; t < \xi, \\ \infty & : \; t \geq \xi. \end{cases}$$

Let $T_K^\beta := \inf_{t>0}\{B^\beta(t) \in K\}$. In analogy to (1.7), we have

Lemma 2.5

$$P_x(T_K^\beta < \infty) = \int_{\mathbb{R}^3} g^\beta((-x) \cdot y) \pi_K^\beta(dy).$$

(Cf. Chaleyat-Maurel, Le Gall (1989), (2.b).) For $\delta > 0$ let $C(\delta)$ be the set of all $\psi \in C^1([0,1], \mathbb{R}^3)$ such that

$$\psi'(t) = a'E'(\psi(t)) + a''E''(\psi(t)) + a'''E'''(\psi(t))$$

for some $a', a'', a''' \in \mathbb{R}$ with $(||(a', a'')||^4 + 16a'''^2)^{1/4} \leq \delta^4$. Define on $\mathbb{R}^3$ the pseudo-distance d given by

$$d(x,y) := \inf\{\delta > 0 : \text{there exists } \psi \in C(\delta) \text{ with } \psi(0) = x, \psi(1) = y\} \wedge 1$$

(cf. Chaleyat-Maurel, Le Gall (1989), p. 225). This pseudo-distance is symmetric, satisfies a generalized triangle inequality on R, and can be estimated by $||.||$ on R:

Lemma 2.6 *(i)*

$$d(x,y) = d(y,x) \quad (x, y \in R),$$

(ii) there is a constant $C > 0$ such that

$$d(x,z) \leq C(d(x,y) + d(y,z)) \quad (x, y, z \in R),$$

(iii) there are constants $C_1, C_2 > 0$ such that

$$C_1 ||x - y|| \leq d(x,y) \leq C_2 ||x - y||^{1/2} \quad (x, y \in R).$$

(Cf. Chaleyat-Maurel, Le Gall (1989), p. 225.)
Furthermore we have:

Lemma 2.7 *There is a constant $D > 0$ such that*
(i)

$$g^0((-x)\cdot y) \le \frac{D}{d^2(x,y)} \qquad (x, y \in \overline{R}).$$

(ii)

$$\lambda(\{y \in \mathbb{R}^3 : d(x,y) < \varepsilon\}) \ge D\varepsilon^4 \qquad \varepsilon > 0 \textit{ small enough.}$$

(Cf. Chaleyat-Maurel, Le Gall (1989), p. 231.)
The next proposition is Chaleyat-Maurel, Le Gall (1989), Proposition 7.1:

Proposition 2.1 *Assume, for some $n \ge 1$, that $x_0, x_1, \dots, x_n$ are distinct points in R. Then we have*

$$\begin{aligned}(\frac{\log(1/\varepsilon)}{\varepsilon})^n P_{x_0}(T_{x_1+\varepsilon K} \le T_{x_2+\varepsilon K} \le \dots \le T_{x_n+\varepsilon K} \le t) \\ \to \quad V_{x_1}(K)V_{x_2}(K)\dots V_{x_n}(K) \int\limits_{0<s_1\le s_2 \le \dots \le s_n < t} p_{s_1}((-x_0)\cdot x_1)p_{s_2-s_1}((-x_1)\cdot x_2)\cdot \dots \\ \cdot p_{s_n-s_{n-1}}((-x_{n-1})\cdot x_n)ds_1 ds_2 \dots ds_n \quad (\varepsilon \to 0) \quad (t \ge 0).\end{aligned} \tag{2.24}$$

There exist constants $C_{n,t}, \overline{C} > 0$ which do not depend on $x_0, x_1, \dots, x_n$ such that, if

$$d(x_{i-1}, x_i) \ge \overline{C}\sqrt{\varepsilon} \quad (i = 1, 2, \dots, n);$$

then the expression on the left hand side of (2.24) can be estimated from above by $C_{n,t}\prod_{i=1}^n d(x_{i-1}, x_i)^{-2}$.

Proof: Let ξ be a random variable obeying to an exponential distribution with parameter β large, independent of $\{B(t)\}_{t\ge 0}$. By Lemma 2.5, for every $x \in R\backslash\{x_0\}$ and $\varepsilon > 0$ small enough, we have

$$P_{x_0}(T_{x+\varepsilon K} < \xi) = \int\limits_{\mathbb{R}^3} g^\beta((-x_0)\cdot y)\pi^\beta_{x+\varepsilon K}(dy). \tag{2.25}$$

By (2.25) and Theorem 2.9 it follows that

$$\frac{\log(1/\varepsilon)}{\varepsilon} P_{x_0}(T_{x+\varepsilon K} < \xi) \to V_x(K)g^\beta((-x_0)\cdot x) \quad (\varepsilon \to 0) \tag{2.26}$$

uniformly for (x_0, x) in a compact subset of $R^2\backslash\{(y,y) : y \in R\}$. There is a constant $C > 0$ such that if $y \in \varepsilon K$, then $d(y, 0) < C\sqrt{\varepsilon}$. By Lemma 2.6 (ii), (iii) there are two constants $c, \overline{C} > 0$ such that for $\varepsilon > 0$ small enough, we have

$$d(x', y') \ge c d(x,y) \quad (x, y \in R, d(x,y) \ge \overline{C}\sqrt{\varepsilon}, x' \in x + \varepsilon K, y' \in y + \varepsilon K). \tag{2.27}$$

It follows from (2.25), Lemma 2.7 (i), and Theorem 2.9 that under the same conditions as in (2.27), for $\varepsilon > 0$ small enough, it holds that

$$\frac{\log(1/\varepsilon)}{\varepsilon} P_{x'}(T_{y+\varepsilon K} < \xi) \leq C'd(x,y)^{-2} \tag{2.28}$$

(if R is not compact, then consider a neighborhood of $\overline{R}$ in order to apply Theorem 2.9). Now we want to prove the second assertion of Proposition 2.1 first. As

$$\begin{aligned} &P_{x_0}(T_{x_1+\varepsilon K} \leq T_{x_2+\varepsilon K} \leq \dots \leq T_{x_n+\varepsilon K} \leq t)P(t \leq \xi) \\ \leq\ &P_{x_0}(T_{x_1+\varepsilon K} \leq T_{x_2+\varepsilon K} \leq \dots \leq T_{x_n+\varepsilon K} < \xi), \end{aligned}$$

it suffices to prove the desired bound with t replaced by ξ. We do induction on n. For $n = 1$ the assertion follows from (2.28). For the induction step, observe that by (2.28), the strong Markov property at time $T_{x_{n-1}+\varepsilon K}$, and the hypothesis $d(x_{n-1}, x_n) \geq \overline{C}\sqrt{\varepsilon}$ we have

$$\begin{aligned} &(\frac{\log(1/\varepsilon)}{\varepsilon})^n P_{x_0}(T_{x_1+\varepsilon K} \leq T_{x_2+\varepsilon K} \leq \dots \leq T_{x_n+\varepsilon K} < \xi) \\ \leq\ &(\frac{\log(1/\varepsilon)}{\varepsilon})^n P_{x_0}(T_{x_1+\varepsilon K} \leq T_{x_2+\varepsilon K} \leq \dots \leq T_{x_{n-1}+\varepsilon K} < \xi, x_n \in S_{\varepsilon K}(T_{x_{n-1}+\varepsilon K}, \xi)) \\ =\ &(\frac{\log(1/\varepsilon)}{\varepsilon})^n E_{x_0}(\mathbf{1}\{T_{x_1+\varepsilon K} \leq T_{x_2+\varepsilon K} \leq \dots \leq T_{x_{n-1}+\varepsilon K} < \xi\} \\ &\cdot E_{B(T_{x_{n-1}+\varepsilon K})}(\mathbf{1}\{T_{x_n+\varepsilon K} < \xi\})) \\ \leq\ &C'd(x_{n-1}, x_n)^{-2}(\frac{\log(1/\varepsilon)}{\varepsilon})^{n-1} P_{x_0}(T_{x_1+\varepsilon K} \leq T_{x_2+\varepsilon K} \leq \dots \leq T_{x_{n-1}+\varepsilon K} < \xi). \end{aligned}$$

Now we want to prove the first assertion. Again we do induction on n. For $n = 1$, define the measure $\nu_\varepsilon^{(1)}(dt)$ on $[0, \infty[$ by

$$\nu_\varepsilon^{(1)}([0,t]) := \frac{\log(1/\varepsilon)}{\varepsilon} P_{x_0}(T_{x_1+\varepsilon K} \leq t).$$

By (2.26), there exists a $\beta_0 > 0$ such that for $\beta \geq \beta_0$ we have

$$\int_0^\infty e^{-\beta t}\nu_\varepsilon^{(1)}(dt) \to V_{x_1}(K)\int_0^\infty e^{-\beta t}p_t((-x_0)\cdot x_1)dt \quad (\varepsilon \to 0). \tag{2.29}$$

(2.29) implies that the measures $\{\nu_\varepsilon^{(1)}\}_{\varepsilon>0}$ converge vaguely, as $\varepsilon \to 0$, to the measure with density

$$V_{x_1}(K)p_t((-x_0)\cdot x_1).$$

This proves the assertion for $n = 1$. For general n, define $\nu_\varepsilon^{(n)}(dt)$ by

$$\nu_\varepsilon^{(n)}([0,t]) := (\frac{\log(1/\varepsilon)}{\varepsilon})^n P_{x_0}(T_{x_1+\varepsilon K} \leq T_{x_2+\varepsilon K} \leq \dots \leq T_{x_n+\varepsilon K} \leq t).$$

By the previous arguments, it suffices to prove that for $\beta \geq \beta_0$ we have

$$\int_0^\infty e^{-\beta t}\nu_\varepsilon^{(n)}(dt)$$
$$\to \prod_{i=1}^n V_{x_i}(K)\int_0^\infty e^{-\beta t}\Big(\int_{0\le s_1<s_2<\ldots<s_{n-1}<t}\Big(\prod_{i=1}^{n-1} p_{s_i-s_{i-1}}((-x_{i-1})\cdot x_i)\Big)$$
$$p_{t-s_{n-1}}((-x_{n-1})\cdot x_n)ds_1ds_2\ldots ds_{n-1}\Big)dt \quad (\varepsilon\to 0),$$

or, equivalently,

$$\Big(\frac{\log(1/\varepsilon)}{\varepsilon}\Big)^n P_{x_0}(T_{x_1+\varepsilon K}\le T_{x_2+\varepsilon K}\le\ldots\le T_{x_n+\varepsilon K}<\xi)$$
$$\to \prod_{i=1}^n (V_{x_i}(K)g^\beta((-x_{i-1})\cdot x_i)). \tag{2.30}$$

So let us verify (2.30). Analogously as above, write

$$\begin{aligned} &P_{x_0}(T_{x_1+\varepsilon K}\le T_{x_2+\varepsilon K}\le\ldots\le T_{x_n+\varepsilon K}<\xi)\\ -\ &P_{x_0}(T_{x_1+\varepsilon K}\le T_{x_2+\varepsilon K}\le\ldots\le T_{x_{n-1}+\varepsilon K}<\xi,\\ &x_n\in S_{\varepsilon K}(T_{x_{n-1}+\varepsilon K},\xi))\\ -\ &P_{x_0}(T_{x_1+\varepsilon K}\le T_{x_2+\varepsilon K}\le\ldots\le T_{x_{n-1}+\varepsilon K}<\xi, T_{x_n+\varepsilon K}<T_{x_{n-1}+\varepsilon K},\\ &x_n\in S_{\varepsilon K}(T_{x_{n-1}+\varepsilon K},\xi)).\end{aligned}\tag{2.31}$$

By the strong Markov property, we get

$$\begin{aligned} &P_{x_0}(T_{x_1+\varepsilon K}\le T_{x_2+\varepsilon K}\le\ldots\le T_{x_{n-1}+\varepsilon K}<\xi, x_n\in S_{\varepsilon K}(T_{x_{n-1}+\varepsilon K},\xi))\\ =\ &E_{x_0}(\mathbf{1}\{T_{x_1+\varepsilon K}\le T_{x_2+\varepsilon K}\le\ldots\le T_{x_{n-1}+\varepsilon K}<\xi\}E_{B(T_{x_{n-1}+\varepsilon K})}(\mathbf{1}\{T_{x_n+\varepsilon K}<\xi\})),\end{aligned}$$

which, by (2.26), behaves like

$$\frac{\varepsilon}{\log(1/\varepsilon)}V_{x_n}(K)g^\beta((-x_{n-1})\cdot x_n)P_{x_0}(T_{x_1+\varepsilon K}\le T_{x_2+\varepsilon K}\le\ldots\le T_{x_{n-1}+\varepsilon K}<\xi).$$

Now (2.30) can easily been proved by induction if we can show that the second term on the right hand side of (2.31) vanishes asymptotically. Indeed, by the Markov property this term is bounded, for $\varepsilon>0$ small enough, by

$$\begin{aligned} &const.\cdot\frac{\varepsilon}{\log(1/\varepsilon)}P_{x_0}(T_{x_1+\varepsilon K}\le T_{x_2+\varepsilon K}\le\ldots\\ &\qquad\le T_{x_{n-1}+\varepsilon K}<\xi, T_{x_n+\varepsilon K}\le T_{x_{n-1}+\varepsilon K})\\ \le\ &const.\cdot\Big(\frac{\varepsilon}{\log(1/\varepsilon)}\Big)^{n+1}\end{aligned}$$

by considering the different possible orderings of $T_{x_i+\varepsilon K}$ $(1\le i\le n)$ and using the (already proved) second assertion of Proposition 2.1. □

Now we are ready for the proof of Theorem 2.8:

Proof of Theorem 2.8: First, we will prove

$$\frac{\log(1/\varepsilon)}{\varepsilon}\kappa(S_{\varepsilon K}(0,t)) \xrightarrow{L^2} \int_0^t \varphi(x_0\cdot B(u))V_{x_0\cdot B(u)}(K)du \quad (\varepsilon\to 0) \quad (t\geq 0). \tag{2.32}$$

It suffices to prove

$$\begin{aligned}
&(\frac{\log(1/\varepsilon)}{\varepsilon})^2 E_{x_0}(\kappa(S_{\varepsilon K}(0,t))^2)\\
\to\ & E_{x_0}(\int_0^t \varphi(x_0\cdot B(u))V_{x_0\cdot B(u)}(K)du)^2\\
=\ & 2\int_R\int_R \varphi(y)\varphi(z)V_y(K)V_z(K)\\
&\int_{0\leq u<v<t} p_u((-x_0)\cdot y)p_{v-u}((-y)\cdot z)du\,dv\,dy\,dz \quad (\varepsilon\to 0)
\end{aligned} \tag{2.33}$$

and

$$\begin{aligned}
&\frac{\log(1/\varepsilon)}{\varepsilon}E_{x_0}(\kappa(S_{\varepsilon K}(0,t))\int_0^t \varphi(x_0\cdot B(u))V_{x_0\cdot B(u)}(K)du)\\
\to\ & E_{x_0}(\int_0^t \varphi(x_0\cdot B(u))V_{x_0\cdot B(u)}(K)du)^2 \quad (\varepsilon\to 0).
\end{aligned} \tag{2.34}$$

We first prove (2.33). We have

$$\begin{aligned}
&(\frac{\log/1/\varepsilon)}{\varepsilon})^2 E_{x_0}(\kappa(S_{\varepsilon K}(0,t))^2)\\
=\ & \int_R\int_R \varphi(y)\varphi(z)(\frac{\log(1/\varepsilon)}{\varepsilon})^2 P_{x_0}(T_{y-\varepsilon K}\leq t, T_{z-\varepsilon K}\leq t)dy\,dz.
\end{aligned} \tag{2.35}$$

Then

$$\begin{aligned}
&P_{x_0}(T_{y-\varepsilon K}\leq t, T_{z-\varepsilon K}\leq t)\\
=\ & P_{x_0}(T_{y-\varepsilon K}\leq T_{z-\varepsilon K}\leq t) + P_{x_0}(T_{z-\varepsilon K}< T_{y-\varepsilon K}\leq t).
\end{aligned}$$

hence by Proposition 2.1

$$\begin{aligned}
&(\frac{\log(1/\varepsilon)}{\varepsilon})^2 P_{x_0}(T_{y-\varepsilon K}\leq t, T_{z-\varepsilon K}\leq t)\\
\to\ & V_y(K)V_z(K)\int_{0\leq u<v<t}(p_u((-x_0)\cdot y))p_{v-u}((-y)\cdot z)\\
&+p_u((-x_0)\cdot z)p_{v-u}((-z)\cdot y)) \quad (\varepsilon\to 0) \quad (y\neq z; y,z\in R\backslash\{x_0\})
\end{aligned}$$

Now we want to calculate the limit of the right hand side of (2.35). On the set $\{d(x_0, y), d(x_0, z), d(y, z) > \overline{C}\sqrt{\varepsilon}\}$ we may, by Proposition 2.1, use the Lebesgue dominated convergence theorem (since $z \mapsto d(y, z)^{-2}$ is integrable on R, cf. Chaleyat-Maurel, Le Gall (1989), p. 254 top). The integral on the complementary set converges to 0, e.g. by Proposition 2.1 and Lemma 2.7 (ii)

$$\int_{d(y,z)<\overline{C}\sqrt{\varepsilon}} (\frac{\log(1/\varepsilon)}{\varepsilon})^2 P_{x_0}(T_{y-\varepsilon K} \leq t, T_{z-\varepsilon K} \leq t) dy\, dz$$

$$\leq const \cdot \varepsilon^2 \int_{\mathbb{R}^3} (\frac{\log(1/\varepsilon)}{\varepsilon})^2 P_{x_0}(T_{y-\varepsilon K} \leq t) dy$$

$$\leq const \cdot (\log(1/\varepsilon))^2 (\int_{d(x_0,y)>\overline{C}\sqrt{\varepsilon}} P_{x_0}(T_{y-\varepsilon K} \leq t) + const \cdot \varepsilon^2)$$

$$\leq const \cdot (\log(1/\varepsilon))^2 (\frac{\varepsilon}{\log(1/\varepsilon)} + const \cdot \varepsilon^2).$$

Now we prove (2.34). We have

$$\frac{\log(1/\varepsilon)}{\varepsilon} E_{x_0}(\kappa(S_{\varepsilon K}(0,t)) \int_0^t \varphi(x_0 \cdot B(u)) V_{x_0 \cdot B(u)}(K) du)$$

$$= \int_R \varphi(y) \frac{\log(1/\varepsilon)}{\varepsilon} E_{x_0}(\mathbf{1}\{T_{y-\varepsilon K} \leq t\} \int_0^t \varphi(x_0 \cdot B(u)) V_{x_0 \cdot B(u)}(K) du) dy$$

$$= \int_R \varphi(y) \frac{\log(1/\varepsilon)}{\varepsilon} (E_{x_0}(\mathbf{1}\{T_{y-\varepsilon K} \leq t\} \int_{T_{y-\varepsilon K}}^t \varphi(x_0 \cdot B(u)) V_{x_0 \cdot B(u)}(K) du)$$

$$+ E_{x_0}(\int_0^t \mathbf{1}\{u \leq T_{y-\varepsilon K} \leq t\} \varphi(x_0 \cdot B(u)) V_{x_0 \cdot B(u)}(K) du)) dy. \tag{2.36}$$

The two terms on the right hand side of (2.36) will be considered separately. Let us first investigate the first one. By the strong Markov property at $T_{y-\varepsilon K}$ we get

$$\frac{\log(1/\varepsilon)}{\varepsilon} E_{x_0}(\mathbf{1}\{T_{y-\varepsilon K} \leq t\} \int_{T_{y-\varepsilon K}}^t \varphi(x_0 \cdot B(u)) V_{x_0 \cdot B(u)}(K) du)$$

$$= \frac{\log(1/\varepsilon)}{\varepsilon} E_{x_0}(\mathbf{1}\{T_{y-\varepsilon K} < t\}$$

$$E_{B(T_{y-\varepsilon K})}(\int_0^{t-T_{y-\varepsilon K}} \varphi(x_0 \cdot B(u)) V_{x_0 \cdot B(u)}(K) du)). \tag{2.37}$$

By the continuity of $y \mapsto p_u((-y) \cdot z)$ for $|z-y| \geq \delta$ and the estimates on the Green function $g^0((-x) \cdot y)$ we get, uniformly for $s \in [0, t]$,

$$E_{y'}(\int_0^{t-s} \varphi(x_0 \cdot B(u)) V_{x_0 \cdot B(u)} du)$$

$$
\begin{aligned}
&= \int_0^t \int_R^s p_u((-y')+z)\varphi(z)V_z(K)dz\,du \\
&\to \int_0^{t\ s} p_u((-y)+z)\varphi(z)V_z(K)dz\,du \qquad (y' \to y).
\end{aligned}
$$

It then follows from (2.37) and Proposition 2.1 that

$$
\begin{aligned}
&\lim_{\varepsilon\to 0} \frac{\log(1/\varepsilon)}{\varepsilon} E_{x_0}(\mathbf{1}\{T_{y-\varepsilon K} \le t\} \int_{T_{y-\varepsilon K}}^{t} \varphi(x_0+B(u))V_{x_0+B(u)}(K)du) \\
&= \lim_{\varepsilon\to 0} \frac{\log(1/\varepsilon)}{\varepsilon} E_{x_0}(\mathbf{1}\{T_{y-\varepsilon K} \le t\} \int_0^{t-T_{y-\varepsilon K}} \int_R p_u((-y)+z)\varphi(z)V_z(K)dz\,du) \\
&= V_y(K)\int_0^t p_s((-x_0)+y) \int_0^s \int_R p_u((-y)+z)\varphi(z)V_z(K)dz\,du\,ds.
\end{aligned}
$$

By integrating with respect to κ and dominated convergence (using the same arguments as above) we get

$$
\begin{aligned}
&\lim_{\varepsilon\to 0} \frac{\log(1/\varepsilon)}{\varepsilon} \int_R \varphi(y) E_{x_0}(\mathbf{1}\{T_{y-\varepsilon K} \le t\} \int_{T_{y-\varepsilon K}}^{t} \varphi(x_0+B(u))V_{x_0+B(u)}(K)du)dy \\
&\to \int_R \int_R \varphi(y)\varphi(z)V_y(K)V_z(K) \\
&\qquad \int_{0\le u\le v\le t} p_u((-x_0)+y)p_{v-u}((-y)+z)du\,dv\,dz\,dy \qquad (\varepsilon \to 0) \qquad (2.38)
\end{aligned}
$$

For the second term on the right hand side of (2.36), write

$$
\begin{aligned}
&E_{x_0}(\mathbf{1}\{u \le T_{y-\varepsilon K} \le t\}\ldots) \\
&= E(\mathbf{1}\{y \in S_{\varepsilon K}(u,t)\}\ldots) - E(\mathbf{1}\{T_{y-\varepsilon K} < u, y \in S_{\varepsilon K}(u,t)\}\ldots).
\end{aligned}
$$

By the fact that

$$
\begin{aligned}
&\frac{\log(1/\varepsilon)}{\varepsilon} E_y(\kappa(S_{\varepsilon K}(0,t-u))) \\
&\to E_y \int_0^{t-u} \varphi((-x_0+B(v))V_{x_0+B(v)}(K)dv \qquad (\varepsilon \to 0) \qquad (y \in R, u \in [0,t])
\end{aligned}
$$

and Proposition 2.1 we obtain (similarly as in the proof of (2.33))

$$
\int_R \varphi(y) \frac{\log(1/\varepsilon)}{\varepsilon} E_{x_0}(\int_0^t \mathbf{1}\{y \in S_{\varepsilon K}(u,t)\}\varphi(x_0+B(u))V_{x_0+B(u)}(K)du)dy
$$

$$= E_{x_0}\Big(\int_0^t \varphi(x_0 + B(u)) V_{x_0+B(u)}(K) \frac{\log(1/\varepsilon)}{\varepsilon} \kappa(S_{\varepsilon K}(u,t))du\Big)$$

$$= E_{x_0}\Big(\int_0^t \varphi(x_0 + B(u)) V_{x_0+B(u)}(K) \frac{\log(1/\varepsilon)}{\varepsilon} E_{x_0+B(u)}(\kappa(S_{\varepsilon K}(0,t-u)))du\Big)$$

$$\to E_{x_0}\Big(\int_0^t \varphi(x_0 + B(u)) V_{x_0+B(u)}(K) E_{x_0+B(u)}\Big(\int_0^{t-u} \varphi(x_0 + B(v)) V_{x_0+B(v)}(K)dv\Big)du\Big)$$

$$= \int_R\int_R \varphi(y)\varphi(z) V_y(K) V_z(K)$$

$$\int_{0<u<v<t} p_u((-x_0)+y)p_{v-u}((-y)+z)du\,dv\,dz\,dy \qquad (\varepsilon \to 0). \tag{2.39}$$

Now the assertion for the second term on the right hand side of (2.36) follows by Proposition 2.1 and Lebesgue's dominated convergence theorem. We see that (2.34) follows from (2.38) and (2.39) if we can prove

$$\frac{\log(1/\varepsilon)}{\varepsilon} \int_R \varphi(y) E_{x_0}\Big(\int_0^t 1\{T_{y-\varepsilon K} < u, y \in S_{\varepsilon K}(u,t)\}$$

$$\varphi(x_0 + B(u)) V_{x_0+B(u)}(K)du\Big)dy \to 0 \quad (\varepsilon \to 0). \tag{2.40}$$

Since φ and $V_{\cdot}(K)$ are bounded, it suffices to prove

$$\frac{\log(1/\varepsilon)}{\varepsilon} E_{x_0}(\lambda(S_{\varepsilon K}(0,u) \cap S_{\varepsilon K}(u,t))) \to 0 \quad (\varepsilon \to 0) \quad (u \in [0,t]). \tag{2.41}$$

Let $\delta > 0$ be small enough. Then

$$\begin{aligned} E_{x_0}(\lambda(S_{\varepsilon K}(0,u) \cap S_{\varepsilon K}(u,u+\delta))) &\le E_{x_0}(\lambda(S_{\varepsilon K}(u,u+\delta))) \\ &\le E_{x_0}(1\{u < \tau\} E_{x_0+B(u)}(\lambda(S_{\varepsilon K}(0,\delta)))). \end{aligned}$$

hence

$$\begin{aligned} &\limsup_{\varepsilon \to 0} \frac{\log(1/\varepsilon)}{\varepsilon} E_{x_0}(\lambda(S_{\varepsilon K}(0,u) \cap S_{\varepsilon K}(0,u+\delta))) \\ &\le E_{x_0}\Big(1\{u < \tau\} E_{x_0+B(u)}\Big(\int_0^\delta V_{x_0+B(v)}(K)dv\Big)\Big) \\ &\le const. \cdot \delta. \end{aligned}$$

Furthermore, for $\delta \le t-u$, since p_δ is bounded and by Proposition 2.1 we get

$$E_{x_0}(\lambda(S_{\varepsilon K}(0,u) \cap S_{\varepsilon K}(u+\delta,t)))$$

$$= E_{x_0}\Big(\int_R 1\{y \in S_{\varepsilon K}(0,u)\} P_{x_0+B(u+\delta)}(y \in S_{\varepsilon K}(0,t-u-\delta))dy\Big)$$

$$- E_{x_0}(\int_R 1\{y \in S_{\epsilon K}(0,u)\} p_\delta((-x_0 \cdot B(u)) \cdot z) P_z(y \in S_{\epsilon K}(0, t-u-\delta)) dy)$$

$$\leq C_\delta \frac{\epsilon}{\log(1/\epsilon)} E_{x_0}(\lambda(S_{\epsilon K}(0,u)))$$

$$\leq C'_\delta (\frac{\epsilon}{\log(1/\epsilon)})^2.$$

Now (2.41) follows readily. This proves (2.34) and (2.32). In fact, the convergence in (2.32) hold in L^p for every $p \in [1, \infty[$ (as in the proof of (2.33), Proposition 2.1 yields that

$$(\frac{\log(1/\epsilon)}{\epsilon})^p E_{x_0}(\kappa(S_{\epsilon K}(0,t))^p) \quad (0 < \epsilon \leq 1)$$

is bounded). Now the rest of the proof is based on an argument in Sznitman (1987), p. 17 (cf. Chaleyat-Maurel, Le Gall (1989), p. 257): The expression

$$E_{x_0}(\sup_{0<s<t<T} |\frac{\log(1/\epsilon)}{\epsilon} \kappa(S_{\epsilon K}(s,t)) - \int_s^t \varphi(x_0 \cdot B(u)) V_{x_0 \cdot B(u)}(K) du|^p)$$

can be approximated by

$$E_{x_0}(\sup_{0<i<j<N} |\frac{\log(1/\epsilon)}{\epsilon} \kappa(S_{\epsilon K}(\frac{iT}{N}, \frac{jT}{N})) - \int_{iT/N}^{jT/N} \varphi(x_0 \cdot B(u)) V_{x_0 \cdot B(u)}(K) du|^p) \qquad (2.42)$$

with N large. Now (2.42) tends to 0 for $\epsilon \to 0$ by (2.32) and the fact that for $0 \leq s \leq t \leq 1$ it holds that (by (2.32) and (2.41))

$$\lim_{\epsilon \to 0} \frac{\log(1/\epsilon)}{\epsilon} \kappa(S_{\epsilon K}(s,t))$$

$$= \lim_{\epsilon \to 0} \frac{\log(1/\epsilon)}{\epsilon} (\kappa(S_{\epsilon K}(0,t)) - \kappa(S_{\epsilon K}(0,s)) + \kappa(S_{\epsilon K}(0,s) \cap S_{\epsilon K}(s,t)))$$

$$= \int_s^t \varphi(x_0 \cdot B(u)) V_{x_0 \cdot B(u)}(K) du. \square$$

The following corollary to Theorem 2.7 (cf. Chaleyat-Maurel, Le Gall (1989), Theorem 7.3) says, roughly speaking, that if small killing sets are thrown randomly into $\mathbb{H}$, then the killed Brownian motion behaves asymptotically like the original Brownian motion $\{x \cdot B(t)\}_{t \geq 0}$ on $\mathbb{H}$ killed at the points $y \in \mathbb{H}$ with a certain rate.
Let $\{a_n\}_{n \geq 1}$ be i.i.d. R-valued random variables obeying to the distribution with bounded density $\varphi(x)$ $(x \in R)$. Assume that the process $\{a_n\}_{n \geq 1}$ is independent of the Brownian motion $\{B(t)\}_{t \geq 0}$ and put

$$R_n := R \setminus \bigcup_{i=1}^n (a_i + \epsilon_n K),$$

$$\tau_n := \inf\{t \geq 0 : B(t) \notin R_n\}.$$

where $\{\varepsilon_n\}_{n\geq 1} \subset]0, \infty[$ is decreasing and such that

$$\lim_{n\to\infty} \frac{n\varepsilon_n}{\log n} = \alpha \in [0, \infty[.$$

Let F be any bounded measurable real-valued function on the Banach space of continuous functions from $[0, t]$ to $I\!H \cong I\!R^3$.

Corollary 2.1 *For every $x \in I\!H$ we have that*

$$E(F(\{x \cdot B(s)\}_{0\leq s\leq t} \cdot 1\{t < \tau_n\}))$$

$$\to \quad E(F(\{x \cdot B(s)\}_{0\leq s\leq t} \cdot 1\{t < \tau\} \exp(-\alpha \int_0^t \varphi(x \cdot B(s)) V_{x\cdot B(s)}(K) ds))) \quad (n \to \infty).$$

The quantity $\alpha\varphi(y)V_y(K)$ may be interpreted as the killing rate at the point y of $\{x \cdot B(s)\}_{0\leq s\leq t}$. The corresponding assumption that has to be made for Brownian motion on $I\!R^3$ is $\lim_{n\to\infty} n\varepsilon_n = \alpha$, which means that the killing sets in our situation have to be larger than in the euclidean case (cf. Chaleyat-Maurel, Le Gall (1989), pp. 258f.).

2.2.2 Transience, recurrence, and the Lebesgue needle

In this section, we continue to present the theory developed by Gallardo (1982). The prerequisites established in 1.3 will be used to formulate an analogue of the classical theorem concerning the "immediate" entrance of Brownian motion into a Lebesgue needle and its transience or recurrence. Notation is as in 1.3. In particular, $|.|$ is the homogeneous norm (1.6).

First, one has to estimate the capacity of a cylinder parallel to the center of $I\!H$. Let $S_L^a \subset I\!H$ be the cylinder

$$S_L^a := \{x = (x', x'', x''') \in I\!H : x'^2 + x''^2 \leq 1, a \leq x''' \leq L + a\},$$

$S_L := S_L^0$.

Proposition 2.2 *Let $L_0 > 0$. Then there exist $M, N > 0$ such that*

$$M \frac{L}{\log L} \leq C(S_L) \leq N \frac{L}{\log L} \quad (L \geq L_0).$$

(Cf. Gallardo (1982), Proposition 6.5.)

Proof: 1. We first prove the upper bound. Since

$$V\pi_{S_L}(x) = P_x(T_{S_L} < \infty) \leq 1,$$

we get

$$L \geq \int_0^L V\pi_{S_L}((0,0,u))du = \int_0^L \int_{S_L} |-x + (0,0,u)|^{-2} \pi_{S_L}(dx) du$$

$$= \int_{S_L} \int_0^L |-x + (0,0,u)|^{-2} du \, \pi_{S_L}(dx).$$

So it suffices to prove

$$\int_0^L |-x+(0,0,u)|^{-2}du \geq \frac{\log L}{N}$$

for $x \in S_L$. Define the homogeneous norm

$$|x|_1 := ((x'^2 + x''^2)^2 + x'''^2)^{1/4}.$$

Since all homogeneous norms are equivalent, we may replace $|.|$ by $|.|_1$ and thus get the estimate

$$\begin{aligned}
\int_0^L |-x+(0,0,u)|^{-2}du &\geq A\int_0^L ((x'^2+x''^2)^2+(u-x''')^2)^{-1/2}du \\
&\geq A\int_0^L (1+(u-x''')^2)^{-1/2}du \\
&\geq A\int_0^L (1+u^2)^{-1/2}du \\
&= A\log(L+\sqrt{L^2+1}) \\
&\geq A\log L.
\end{aligned}$$

2. Now we come to the lower bound. Since $L \mapsto \frac{L}{\log L}$ is concave and $L \mapsto C(S_L)$ is increasing, we may w.l.o.g. assume $L \in \mathbb{N}$. We get

$$\begin{aligned}
V\pi_{S_1^j}(x) &= P_x(T_{S_1^j} < \infty) \\
&= P_{(0,0,-j)\cdot x}(T_{(0,0,-j)\cdot S_1^j} < \infty) \\
&= P_{x\cdot(0,0,j)}(T_{S_1} < \infty) \\
&= V\pi_{S_1}(x-(0,0,j)).
\end{aligned}$$

So if we consider the measure

$$\eta_L := \sum_{j=1}^{L} \pi_{S_1^{j-1}}$$

supported by S_L, we have

$$\eta_L(S_L) = L\cdot C(S_1) \tag{2.43}$$

and

$$V\eta_L(x) = \sum_{j=1}^{L} V\pi_{S_1}(x-(0,0,j)). \tag{2.44}$$

Consider the homogeneous norm

$$|x|_\infty := \sup\{\sqrt{x'^2+x''^2}, \sqrt{|x'''|}\}.$$

By Proposition 1.5 and the equivalence of homogeneous norms, we get

$$\begin{aligned} V\pi_{S_1}(x) &\le \left(\frac{D}{|x|_\infty}\right)^2 \wedge 1 \\ &\le \frac{D^2}{|x'''|} \wedge 1 \quad (x \in \mathbb{H}), \end{aligned}$$

hence by (2.44) and an easy calculation

$$\begin{aligned} V\eta_L(x) &\le \sum_{j=1}^{L} \frac{D^2}{|x''' - j|} \wedge 1 \\ &\le K \log L \quad (x \in \mathbb{H}). \end{aligned}$$

So

$$V\frac{\eta_L}{K \log L}(x) \le 1 \quad (x \in \mathbb{H}).$$

and since η_L is supported by S_L, we get by (2.43) and Lemma 1.3

$$C(S_L) \ge \frac{\eta_L(S_L)}{K \log L} = C(S_1)\frac{L}{K \log L}. \square$$

Now we formulate the theorem on the Lebesgue needle. It is a criterion if a Brownian motion starting at 0 enters "immediately" into a set which has a sharper vertex at 0 than a cone. (Cf. Gallardo (1982), Theorem 6.2.)

Theorem 2.10 *Let $h :]0,\infty[\to]0,\infty[$ be a measurable function such that $h(u)/\sqrt{u} \downarrow 0$ $(u \to 0)$,*

$$A_h := \{x = (x', x'', x''') \in \mathbb{H} : \sqrt{x'^2 + x''^2} \le h(x'''), x''' > 0\}.$$

The following conditions are equivalent:
(i) $P_0(T_{A_h} = 0) = 1$,
(ii) $\int_0^1 |\log \frac{h(u)}{\sqrt{u}}|^{-1} \frac{du}{u} = \infty$.

Proof: Let $C_1 \ge 1$ be a constant such that

$$\frac{1}{C_1}|x|_\infty \le |x| \le C_1|x|_\infty \quad (x \in \mathbb{H}).$$

Let $\alpha > C_1^2$. Put

$$A_n := \{x \in A_h : \alpha^{-n-1} < |x| \le \alpha^{-n}\},$$

$$A'_n := \{x \in A_h : C_1\alpha^{-n-1} < |x|_\infty \le \frac{1}{C_1}\alpha^{-n}\},$$

$$A''_n := \{x \in A_h : \frac{1}{C_1}\alpha^{-n-1} < |x|_\infty \le C_1\alpha^{-n}\}.$$

Clearly, $\emptyset \neq A'_n \subset A_n \subset A''_n$. By the assumption on h, it follows that for $n \geq n_0$

$$A'_n = \{x \in A_h : (C_1\alpha^{-(n+1)})^2 < x''' \leq (\frac{\alpha^{-n}}{C_1})^2\},$$

$$A''_n := \{x \in A_h : (\frac{\alpha^{-(n-1)}}{C_1})^2 < x''' \leq (C_1\alpha^{-n})^2\}.$$

Define the cylinders

$$S'_n := \{x \in I\!H : |(x', x'', 0)| \leq h((C_1\alpha^{-(n+1)})^2), (C_1\alpha^{-(n+1)})^2 < x''' \leq (\frac{\alpha^{-n}}{C_1})^2\},$$

$$S''_n := \{x \in I\!H : |(x', x'', 0)| \leq h((C_1\alpha^{-n})^2), (\frac{\alpha^{-(n+1)}}{C_1})^2 < x''' \leq (C_1\alpha^{-n})^2\}.$$

Since $h(u)$ is decreasing, it follows that

$$\emptyset \neq S'_n \subset A'_n \subset A_n \subset A''_n \subset S''_n \quad (n \geq n_0).$$

Now observe that $\delta_r(S'_n)$ is a cylinder of height

$$r^2((\frac{\alpha^{-n}}{C_1})^2 - (C_1\alpha^{-(n+1)})^2) = (r\alpha^{-(n+1)})^2\frac{\alpha^2 - C_1^4}{C_1^2}$$

and radius

$$rh((C_1\alpha^{-(n+1)})^2).$$

Take $r_0 := (h((C_1\alpha^{-(n+1)})^2))^{-1}$. Then by Proposition 2.2

$$\begin{aligned} C(\delta_{r_0}(S'_n)) &= r_0^2 C(S'_n) \\ &\geq M\frac{\alpha^2 - C_1^4}{C_1^2}(r_0\alpha^{-(n+1)})^2(\log(\frac{\alpha^2 - C_1^4}{C_1^2}(r_0\alpha^{-(n+1)})^2))^{-1}, \end{aligned}$$

hence

$$C(S'_n) \geq A(\alpha^{-(n+1)})^2(\log(A'(\frac{\alpha^{-(n+1)}}{h((C_1\alpha^{-(n+1)})^2)})^2))^{-1} \quad (n \geq n_1).$$

By a similar calculation, one gets the upper estimate

$$C(S''_n) \leq B(\alpha^{-n})^2(\log(B'(\frac{\alpha^{-n}}{h((C_1\alpha^{-n})^2)})^2))^{-1} \quad (n \geq n_2).$$

Now by the inclusion $S'_n \subset A_n \subset S''_n$, the fact that $A \subset B \Longrightarrow C(A) \leq C(B)$, Theorem 1.7, and an easy calculation the assertion follows. □

By Theorem 1.8 one can prove in an analogous way as Theorem 2.10 (cf. Gallardo (1982), Theorem 7.4):

Theorem 2.11 *With the notation of Theorem 2.10, but* $\frac{h(u)}{\sqrt{u}} \downarrow 0 \quad (u \to \infty)$*, the following conditions are equivalent:*

(i) P_0(*for any* $t > 0$ *there exists* $s > t$ *such that* $B(s) \in A_h$) $= 1$

(ii) $\int_1^\infty |\log \frac{h(u)}{\sqrt{u}}|^{-1} \frac{du}{u} = \infty$.

For stratified groups G of higher step, the analogues of Theorems 2.10 and 2.11 hold with $\frac{h(u)}{\sqrt{u}} \downarrow 0$ replaced by $\frac{h(u)}{u^{1/r}} \downarrow 0$ and $|\log \frac{h(u)}{\sqrt{u}}|^{-1}$ in (ii) replaced by $(\frac{h(u)}{u^{1/r}})^{k-2-r}$, where k is the homogeneous dimension and r the step of nilpotency of G. The proofs parallel (with some exceptions) those of Theorems 2.10 and 2.11 (cf. Gallardo (1982)).

2.3 A.s. results

2.3.1 Local and asymptotic behavior of Brownian motion

In the classical situation, the following law of the iterated logarithm for $\{B_1(t)\}_{t>0}$ is due to Lévy (cf. Csörgö, Révész (1981), Theorem 1.3.1):

Theorem 2.12

$$\limsup_{t\to\infty} \frac{|B_1(t)|}{\sqrt{2t\log\log t}} \stackrel{a.s.}{=} 1.$$

If one defines

$$\tilde{B}_1 := \begin{cases} tB_1(\frac{1}{t}) & : \ t>0 \\ 0 & : \ t=0, \end{cases}$$

then $\tilde{B}_1$ is again a Gaussian process, and by comparing expectations and covariances one obtains

$$\mathcal{L}(\{\tilde{B}_1(t)\}_{t>0}) = \mathcal{L}(\{B_1(t)\}_{t\geq 0}).$$

Thus the asymptotic law of the iterated logarithm given in Theorem 2.12 is equivalent to the following "local" form (cf. Csörgö, Révész (1981), Theorem 1.3.3):

Theorem 2.13

$$\limsup_{t\to 0} \frac{|B_1(t)|}{\sqrt{2t\log\log\frac{1}{t}}} \stackrel{a.s.}{=} 1.$$

An analogue of Theorem 2.12 for the central component $A(t)$ of $B(t)$ on $I\!H$ (even in a much more general form, which is e.g. also applicable to non-standard Brownian motions) was given by Berthuet (1979, 1986):

Theorem 2.14

$$\limsup_{t\to\infty} \frac{\pi|A(t)|}{t\log\log t} \stackrel{a.s.}{=} 1.$$

See also Baldi (1986) for a qualitative (functional) version of this result on general simply connected nilpotent Lie groups. His proof rests on the Freidlin-Ventsel (1984) theory on random permutations of a dynamical system. However, in this more general case, the analogue of the constant π is not known. Probably, in contrast to the classical case, there is no direct equivalence between the asymptotic and the local form of the law of the iterated logarithm here. Nevertheless, the local form is also true, as was shown by Helmes (1986). However, we will give a proof which is different from the martingale-theoretic one of Helmes (1986). Ours combines the methods in Loève (1978), pp. 249ff. with a "maximal lemma" (cf. Berthuet (1979), Lemma 2) for $\{A(t)\}_{t>0}$:

Lemma 2.8

$$P(\sup_{0\leq t<1} A(t) \geq u) = 2P(A(1)\geq u) \quad (u\in I\!R).$$

Theorem 2.15

$$\limsup_{t\to 0} \frac{\pi\,|A(t)|}{t\log\log\frac{1}{t}} \overset{a.s.}{=} 1.$$

Proof: We have to show

$$\limsup_{t\to 0} \pi A(t)/(t\log\log(1/t)) \overset{a.s.}{\leq} b \quad (b>1), \tag{2.45}$$

$$\limsup_{t\to 0} \pi A(t)/(t\log\log(1/t)) \overset{a.s.}{\geq} c \quad (c<1). \tag{2.46}$$

Put

$$g(t) := t\log\log(1/t).$$

1. We first prove (2.45). Put, for $n \in \mathbb{N}$,

$$t_n := \left(\frac{b+1}{2b}\right)^n$$

and consider the events

$$C_n := \{\pi A(t)/g(t) > b \quad \text{for some } t \in [t_{n+1}, t_n]\}.$$

With the aid of Lemma 2.8 we get, for $n \geq n_0$,

$$\begin{aligned}
P(C_n) &\leq P(\pi \sup_{0\leq s\leq t_n} A(s)/g(t_{n+1}) > b) \\
&\leq \frac{1}{c} P(\pi A(t_n)/g(t_{n+1}) > b - \pi t_n/g(t_{n+1})) \\
&\leq \frac{1}{c} P(\pi A(t_n)/g(t_{n+1}) > \frac{3b}{b+2}) \\
&= \frac{1}{c} P(\pi A(1) t_n/g(t_{n+1}) > \frac{3b}{b+2}) \\
&= \frac{1}{c} \int_{\frac{3bg(t_{n+1})}{\pi(b+2)t_n}}^{\infty} (\cosh(\pi x))^{-1}\,dx \\
&= O(\exp(-\frac{3bg(t_{n+1})}{(b+2)t_n})) \\
&= O((n+1)^{-\frac{3(b+1)}{2(b+2)}}) \quad (n\to\infty).
\end{aligned}$$

Since

$$\frac{3(b+1)}{2(b+2)} > 1$$

we get

$$\sum_{n=1}^{\infty} P(C_n) < \infty,$$

which, by the Borel-Cantelli Lemma, implies (2.45).
2. Now we prove (2.46). Put, for $n \in \mathbb{N}$,

$$t_n := w^n,$$

where $w \in]0,1[$ is small enough. It suffices to verify

$$P(\pi A(t_n)/g(t_n) \geq c \quad \text{for infinitely many } n) = 1. \tag{2.47}$$

We get, for suitable $H, K > 0$ and n big enough,

$$\begin{aligned} P(\pi((-B(t_{n+1}))\cdot B(t_n))'''/g(t_n) \geq 1-w) \quad & P(\pi(B(t_n - t_{n-1}))'''/g(t_n) \geq 1-w) \\ &= P(\pi A(t_n - t_{n-1})/g(t_n) \geq 1-w) \\ &= P(\pi \frac{A(t_n - t_{n+1})}{t_n - t_{n-1}} \frac{t_n}{g(t_n)} \geq 1) \\ &= P(\pi A(1)\frac{t_n}{g(t_n)} \geq 1) \\ &= \int_{\frac{g(t_n)}{\pi t_n}}^{\infty} (\cosh(\pi x))^{-1} dx \\ &\geq H \exp(-\frac{g(t_n)}{t_n}) \\ &= K n^{-1}. \end{aligned}$$

So we have

$$\sum_{n=1}^{\infty} P(\pi((-B(t_{n+1}))\cdot B(t_n))'''/g(t_n) \geq 1-w) = \infty.$$

Since the increments $(-B(t_{n+1}))\cdot B(t_n)$ are independent, the Borel-Cantelli Lemma implies

$$P(\pi((-B(t_{n+1}))\cdot B(t_n))'''/g(t_n) \geq 1-w \quad \text{for infinitely many } n) = 1. \tag{2.48}$$

Furthermore we get

$$P(\pi|\frac{1}{2}[B(t_{n+1}), B(t_n) - B(t_{n+1})]'''|/g(t_n) \geq \sqrt[3]{w} \quad \text{for infinitely many } n) = 0 \tag{2.49}$$

also by the Borel-Cantelli Lemma and the fact that w is chosen small enough, for

$$\begin{aligned} P(\pi|\frac{1}{2}[B(t_{n+1}), B(t_n) - B(t_{n+1})]'''|/g(t_n) \geq \sqrt[3]{w}) &= O(\exp(-\frac{K\sqrt[3]{w}g(t_n)}{\sqrt{t_{n+1}(t_n - t_{n+1})}})) \\ &= O(n^{-\frac{K}{\sqrt[6]{w}\sqrt{1-w}}}) \quad (n \to \infty) \end{aligned}$$

for some $K > 0$ which is independent of q (observe that for random variables X_1, X_2, X_3, X_4 each obeying to a standard normal law the following trivial estimation holds:

$$P(|X_1X_2 - X_3X_4| > x) = O(P(X_1 > \sqrt{x/2}))$$

$$
\begin{aligned}
&= O\Big(\int_{\sqrt{x/2}}^{\infty} e^{-t^2/2}dt\Big) \\
&= O\Big(\int_{\sqrt{x/2}}^{\infty} te^{-t^2/2}dt\Big) \\
&= O(e^{-x/4}) \quad (x \to \infty)).
\end{aligned}
$$

Now, by (2.45), (2.48), (2.49), and the symmetry of A we have a.s. for infinitely many n

$$
\begin{aligned}
\pi A(t_n)/g(t_n) &= \pi B(t_n)'''/g(t_n) \\
&\geq \pi((-B(t_{n+1}))\cdot B(t_n))'''/g(t_n) - \pi|A(t_{n+1})|/g(t_n) \\
&\quad -\pi\left|\frac{1}{2}|B(t_{n+1}), B(t_n) - B(t_{n+1})|'''\right|/g(t_n) \\
&\geq 1 - w - 2w - \sqrt[3]{w}.
\end{aligned}
$$

Hence, as w can be chosen as near to 0 as we want, we obtain (2.47) and thus (2.46).□
In the classical literature, there is also the so-called "other" or "maximal" law of the iterated logarithm due to Chung (1948) (see also Bingham (1986), p. 444) available. For Brownian motion $\{B_1(t)\}_{t\geq 0}$ this tells the following:

Theorem 2.16

$$
\liminf_{t\to\infty} \sqrt{\frac{\log\log t}{t}} \sup_{0\leq s\leq t} |B_1(s)| \overset{a.s.}{=} \frac{\pi}{\sqrt{8}}.
$$

An analogue for $\{A(t)\}_{t>0}$ is Shi (1995), Corollary 2 resp. Rémillard (1994), Theorem 1 and the Remark on p. 1799:

Theorem 2.17

$$
\liminf_{t\to\infty} \frac{\log\log t}{t} \sup_{0<s<t} |A(s)| \overset{a.s.}{=} \frac{\pi}{4}.
$$

We will present Shi's proof, which is based on an integral test for $P(\sup_{0\leq s\leq t}|A(s)| < tg(t)$ infinitely often) being 0 or 1 for non-increasing positive functions g such that $t \mapsto tg(t)$ is non-decreasing (cf. Proposition 2.3). Interestingly enough, Shi's (1995) method yields related results for windings of two-dimensional Brownian motion. The proof of Theorem 2.17 will be prepared by some auxiliary results.

Lemma 2.9 *Let $\{B_1(t)\}_{t\geq 0}$ be $\mathbb{R}$-valued standard Brownian motion, $s, t, \alpha, \beta > 0$. Then*

$$
\begin{aligned}
&E\Big(\exp\Big(-\frac{\alpha^2}{2}\int_0^t B_1(u)^2du - \frac{\beta^2}{2}\int_t^{t+s} B_1(u)^2du\Big)\Big) \\
&= \Big(\sinh\alpha t\cdot\cosh\beta s\cdot\Big(\coth\alpha t + \frac{\beta}{\alpha}\tanh\beta s\Big)\Big)^{-1/2}.
\end{aligned}
$$

(Cf. Shi (1995), Lemma 4.) This can be proved by using known results on Laplace transforms of quadratic Brownian functionals or by directly solving the associated Sturm-Liouville equation. Since

$$\sinh \alpha t \cdot (\coth \alpha t + \frac{\beta}{\alpha} \tanh \beta s) \geq \cosh \alpha t$$

and

$$\frac{1}{2} e^{x} \leq \cosh x \leq e^{x} \quad (x \geq 0),$$

we get from Lemma 2.9 the following corollary (cf. Shi (1995), (3.6), (3.7)):

Corollary 2.2 *Let $\{R(t)\}_{t \geq 0}$ be as in (2.5). Then for all $s, t, \alpha, \beta > 0$ we have*
(i)

$$e^{-\alpha t} \leq E(\exp(-\frac{\alpha^2}{2} \int_0^t R(v)^2 dv)) \leq 2e^{-\alpha t},$$

(ii)

$$E(\exp(-\frac{\alpha^2}{2} \int_0^t R(v)^2 dv - \frac{\beta^2}{2} \int_t^{t+s} R(v)^2 dv)) \leq 4e^{-\alpha t - \beta s}.$$

The following estimation is Shi (1995), (2.2), (2.3):

Lemma 2.10 *Let $\{B_1(t)\}_{t \geq 0}$ be $\mathbb{R}$-valued standard Brownian motion and $\{H(t)\}_{t \geq 0}$ a positive non-decreasing continuous process independent of $\{B_1(t)\}_{t \geq 0}$. Then for $0 \leq s \leq t, 0 \leq x \leq y$ we have*
(i)

$$\begin{aligned} \frac{8}{3\pi} E(\exp(-\frac{\pi^2}{8x^2} H(t))) &\leq P(\sup_{0 < u \leq t} |B_1(H(u))| < x) \\ &\leq \frac{4}{\pi} E(\exp(-\frac{-\pi^2}{8x^2} H(t))), \end{aligned}$$

(ii)

$$\begin{aligned} & P(\sup_{0 \leq u \leq s} |B_1(H(u))| < x, \sup_{0 \leq u \leq t} |B_1(H(u))| < y) \\ \leq\ & \frac{16}{\pi^2} E(\exp(-\frac{\pi^2}{8x^2} H(s) - \frac{\pi^2}{8y^2} (H(t) - H(s)))). \end{aligned}$$

Now we mention Shi (1995), Lemma 3:

Lemma 2.11 *Let g be a positive non-increasing function on $]b, \infty[$ such that $t \mapsto tg(t)$ is non-decreasing. Let $t_0 > 0$ be large enough and define the sequence $\{t_k\}_{k \geq 0}$ recursively by*

$$t_{k+1} := (1 + g(t_k)) t_k.$$

Write $g_k := g(t_k)$. Then the expressions

$$\int_{t_0}^{\infty} \exp(-\frac{\pi}{4g(t)}) \frac{dt}{tg(t)} \quad , \quad \sum_{k=0}^{\infty} \exp(-\frac{\pi}{4g_k})$$

converge and diverge simultaneously.

Proof: Since $t \mapsto tg(t)$ is non-decreasing, we have

$$\frac{g_k}{g_{k+1}} \le \frac{t_{k+1}}{t_k} = 1 + g_k, \tag{2.50}$$

which is bounded. Hence

$$\begin{aligned}
&\int_{t_0}^{\infty} \exp(-\frac{\pi}{4g(t)})\frac{dt}{tg(t)} \\
&= \sum_{k=0}^{\infty} \int_{t_k}^{t_{k+1}} \exp(-\frac{\pi}{4g(t)})\frac{dt}{tg(t)} \\
&\le \sum_{k=0}^{\infty} \exp(-\frac{\pi}{4g_k})\frac{1}{g_{k+1}} \log\frac{t_{k+1}}{t_k} \\
&= \sum_{k=0}^{\infty} \exp(-\frac{\pi}{4g_k})\frac{1+g_k}{g_k} \log(1+g_k) \\
&\le C\sum_{k=0}^{\infty} \exp(-\frac{\pi}{4g_k})
\end{aligned}$$

(since $\frac{(1+x)\log(1+x)}{x} \to 1 \quad (x \to 0)$). On the other hand, by (2.50),

$$\begin{aligned}
\int_{t_0}^{\infty} \exp(-\frac{\pi}{4g(t)})\frac{dt}{tg(t)} &\ge \sum_{k=0}^{\infty} \exp(-\frac{\pi}{4g_{k+1}})\frac{1}{g_k}\log(1+g_k) \\
&\ge C'\sum_{k=0}^{\infty} \exp(-\frac{\pi}{4g_{k+1}})
\end{aligned}$$

(since $\frac{\log(1+x)}{x} \to 1 \quad (x \to 0)$). □
Now Theorem 2.17 will be a consequence of the following integral test:

Proposition 2.3 *Let $\{t_k\}_{k\ge 0}$ be the sequence of Lemma 2.11. Let also g be as in Lemma 2.11 with the additional assumption that $tg(t) \uparrow \infty \quad (t \to \infty)$. Then*

$$P(\sup_{0\le u\le t} |A(u)| < tg(t) \textit{ for arbitrary large } t) = \begin{cases} 0 \\ 1 \end{cases}$$

iff

$$\int_{t_0}^{\infty} \frac{dt}{tg(t)} \exp(-\frac{\pi}{4g(t)}) \begin{cases} < \infty \\ = \infty. \end{cases}$$

Proof: 1. Define the events

$$A_k := \{ \sup_{0<u<t_k} |A(u)| < t_{k+1}g_{k+1} \}.$$

By Theorem 2.12, Lemma 2.10 (i), Corollary 2.2 (i), and (2.50) we get

$$\begin{aligned} P(A_k) &\le \frac{4}{\pi} E(\exp(-\frac{\pi^2}{32 t_{k+1}^2 g_{k-1}^2} \int_0^{t_k} R(v)^2 dv)) \\ &\le \frac{8}{\pi} \exp(-\frac{\pi t_k}{4 t_{k+1} g_{k+1}}) \\ &= \frac{8}{\pi} \exp(-\frac{\pi}{4(1+g_k) g_{k+1}}) \\ &= \frac{8}{\pi} \exp(-\frac{\pi}{4 g_{k+1}} + \frac{\pi}{4} \frac{g_k}{g_{k+1}(1+g_k)}) \\ &\le K \exp(-\frac{\pi}{4 g_{k-1}}). \end{aligned}$$

Now Lemma 2.11 together with the Borel-Cantelli lemma and a monotonicity argument yield

$$\int \ldots < \infty \Longrightarrow P(\ldots) = 0$$

of Proposition 2.3.
2. Suppose $\int \ldots = \infty$ in Proposition 2.3. Then by Lemma 2.11

$$\sum_{k=0}^{\infty} \exp(-\frac{\pi}{4 g_k}) < \infty.$$

Define the events

$$D_k := \{ \sup_{0 \le u \le t_k} |A(u)| < t_k g_k \}.$$

By Lemma 2.10 (i) and Corollary 2.2 (i) it follows that

$$\begin{aligned} P(D_k) &\ge \frac{8}{3\pi} E(\exp(-\frac{-\pi^2}{32 t_k^2 g_k^2} \int_0^{t_k} R(v)^2 dv)) \\ &\ge \frac{8}{3\pi} \exp(-\frac{\pi}{g_k}), \end{aligned} \tag{2.51}$$

whereas by Lemma 2.10 (ii) and Corollary 2.2 (ii)

$$\begin{aligned} P(D_k D_\ell) &\le \frac{16}{\pi^2} E(\exp(-\frac{\pi^2}{32 t_k^2 g_k^2} \int_0^{t_k} R(v)^2 dv - \frac{\pi^2}{32 t_\ell^2 g_\ell^2} \int_{t_k}^{t_\ell} R(v)^2 dv)) \\ &\le \frac{64}{\pi^2} \exp(-\frac{\pi}{4 g_k} - \frac{\pi (t_\ell - t_k)}{4 t_\ell g_\ell}) \\ &\le \frac{64}{\pi^2} \exp(-\frac{\pi}{4 g_k}) \exp(\frac{\pi t_k}{4 t_\ell g_\ell}). \end{aligned} \tag{2.52}$$

Now (2.51) and (2.52) yield

$$\sum_{k=1}^{n} \sum_{\ell=1}^{n} P(B_k B_\ell) \le K (\sum_{k=1}^{n} P(B_k))^2,$$

which proves $P(\dots) = 1$ in Proposition 2.3 by Kochen-Stone's version of the Borel-Cantelli lemma (cf. Kochen, Stone (1964)) and the $0-1$-law Lemma 1.4 (the invariance follows from $tg(t) \uparrow \infty \quad (t \to \infty)$). □
Theorem 2.17 is now an immediate consequence of Proposition 2.3. Rémillard's (1994) proof is more operator-theoretic.
Concerning the local behavior of $\{B_1(t)\}_{t\geq 0}$, Lévy and Csörgö, Révész proved also (cf. Csörgö, Révész (1981), Theorem 1.1.1, Theorem 1.6.1):

Theorem 2.18 *(i)*

$$\frac{\sup_{0<s<1-h} |B_1(s+h) - B_1(s)|}{\sqrt{2h \log \frac{1}{h}}} \overset{a.s.}{\to} 1 \quad (h \to 0),$$

(ii)

$$\frac{\sup_{0\leq s\leq 1-h} \sup_{0<t\leq h} |B_1(s+t) - B_1(s)|}{\sqrt{2h \log \frac{1}{h}}} \overset{a.s.}{\to} 1 \quad (h \to 0).$$

(iii)

$$\inf_{0<s<1-h} \sup_{0\leq t\leq h} \sqrt{\frac{8 \log \frac{1}{h}}{\pi^2 h}} |B_1(s+t) - B_1(s)| \overset{a.s.}{\to} 1 \quad (h \to 0).$$

This theorem gives information on the modulus of continuity of $\{B_1(t)\}_{t>0}$. Baldi (1990), Theorem 4.1 is a strong uniform (in (s,t)) limit theorem for the paths

$$[0,1] \ni t \mapsto \delta_{\sqrt{h \log(1/h)}}(B(s+th) \cdot (-B(s)))$$

(as $h \to 0$) for general simply connected nilpotent Lie groups, which implies at once the analogues of Theorem 2.18 (i),(ii) (in qualitative form). The work of Baldi is based on the Freidlin-Ventsel (1984) theory on large deviation estimates for diffusion processes having a small diffusion coefficient. For $I\!H$ one has even the exact quantitative form of (i) and (ii):

Theorem 2.19

$$\frac{\pi \sup_{0\leq s\leq 1-h} |q((-B(s)) \cdot B(s+h))|}{h \log(1/h)} \overset{a.s.}{\to} 1 \quad (h \to 0),$$

$$\frac{\pi \sup_{0\leq s\leq 1-h} \sup_{0<t<h} |q((-B(s)) \cdot B(s+t))|}{h \log(1/h)} \overset{a.s.}{\to} 1 \quad (h \to 0).$$

Proof: By Baldi (1990), p.200, lines 7-9 from below and Baldi (1986), Corollary 3.2 and p.443, lines 7-10 from below, it follows that

$$\begin{aligned} &\lim_{h\to 0} \frac{\sup_{0\leq s\leq 1-h} |q((-B(s)) \cdot B(s+h))|}{h \log(1/h)} \\ &\overset{a.s.}{=} \lim_{h\to 0} \frac{\sup_{0\leq s<1-h} \sup_{0<t<h} |q((-B(s)) \cdot B(s+t))|}{h \log(1/h)} \\ &\overset{a.s.}{=} c, \end{aligned}$$

where c is the constant determined in the proof of Baldi (1986), Corollary 3.3 for $\alpha = -\beta = \frac{1}{2}$. This proof consists of solving a variational problem and runs as follows: It turns out that c is the maximum value of the functional

$$J(f) := \frac{1}{2}\int_0^1 [f_1'(s)f_2(s) - f_1(s)f_2'(s)]ds,$$

where $f = (f_1, f_2)$ runs over the ball with radius 2 of the Hilbert space $\mathcal{H}_2$ of absolutely continuous functions $[0,1] \to I\!R^2$ whose derivative is square integrable and which satisfy $f(0) = 0$, equipped with the norm $|f|_1^2 := \int_0^1 (f_1'(s)^2 + f_2'(s)^2)ds$. The Gateaux differential of J is given by

$$J'(f)(h) = \frac{1}{2}\int_0^1 [(h_1'(s)f_2(s) + f_1'(s)h_2(s)) - (h_1(s)f_2'(s) + f_1(s)h_2'(s))]ds.$$

J' vanishes only for $f = 0$, so the maximum is attained at an f with $|f|_1^2 = 2$ and such that there exists a Lagrangian multiplier $\lambda \in I\!R$ satisfying

$$J'(f)(h) = \lambda \int_0^1 (f_1'(s)h_1'(s) + f_2'(s)h_2'(s))ds \quad (h \in \mathcal{H}_2). \tag{2.53}$$

Since $J'(f)(f) = 2J(f)$ one gets

$$J(f) = \frac{1}{2}J'(f)(f) = \frac{\lambda}{2}|f|_1^2 = \lambda.$$

Now in (2.53) we substitute functions h given by $h_{i_1} = \int \mathbf{1}\{[0,s]\}, h_{i_2} = 0$ $((i_1, i_2) \in \{(1,2),(2,1)\})$, with which we get the system of differential equations

$$f'(s) = \frac{1}{\lambda}\begin{pmatrix} 0 & -1 \\ 1 & 0 \end{pmatrix} f(s) + \frac{1}{2\lambda}\begin{pmatrix} 0 & 1 \\ -1 & 0 \end{pmatrix} v,$$

where $v = f(1)$. This system is equivalent to (2.53), since the set of linear combinations of functions h of the above-mentioned form, as s varies in $[0,1]$, is dense in $\mathcal{H}_2$. The solution of the above system is given by

$$f(s) = (\exp\begin{pmatrix} -1 & -s/\lambda \\ s/\lambda & -1 \end{pmatrix})\begin{pmatrix} 0 & -1 \\ 1 & 0 \end{pmatrix}^{-1}\frac{1}{2}\begin{pmatrix} 0 & 1 \\ -1 & 0 \end{pmatrix} f(1).$$

But $f(1) = v$ iff

$$1 \in \operatorname{Spec}(\exp\begin{pmatrix} -1 & -1/\lambda \\ 1/\lambda & -1 \end{pmatrix})\begin{pmatrix} 0 & -1 \\ 1 & 0 \end{pmatrix}^{-1}\frac{1}{2}\begin{pmatrix} 0 & 1 \\ -1 & 0 \end{pmatrix}),$$

which yields

$$\cos(\frac{1}{\lambda}) = -1,$$

i.e. $c = J(f) = \lambda - \frac{1}{\pi}$. $\square$

The following theorem (cf. Neuenschwander, Schott (1995), Theorem 1) is a qualitative form of an analogue of the Erdös-Rényi law of large numbers for Brownian motion (cf. Csörgö, Révész (1981), Corollary 1.2.1). For functions f, g, the symbol $f = \Theta(g)$ means that both $f = O(g)$ and $g = O(f)$.

Theorem 2.20 *For any $c > 0$ we have*

$$\sup_{0 \le t \le T - c\log T} \frac{|q((-B(t)) \cdot B(t + c\log T))|}{(\log T)^2} = \Theta(1) \qquad (T \to \infty).$$

For the proof we need some preparation. Let $|.|$ be the homogeneous norm (1.1) on $I\!H$. The following lemma is the analogue of Lemma 1.1.1 in Csörgö, Révész (1981) (see also Neuenschwander, Schott (1995), Lemma 2):

Lemma 2.12 *There is a constant H and, for every $\varepsilon > 0$, a constant $D = D(\varepsilon)$ such that for every $v > 0, 0 < h < 1$*

$$P\Big(\sup_{0 \le s \le 1-h} \sup_{0 < t \le h} |q((-B(s)) \cdot B(s+t))| \ge vh\Big) \le \frac{D}{h} \exp\Big[-\frac{v}{H+\varepsilon}\Big].$$

Proof: W.l.o.g. we may assume $v \ge 1$ (say), for otherwise the lemma is trivial. Let $\varepsilon > 0$ be fixed. For $s > 0$ let

$$s = \sum_{j=0}^{\infty} \frac{\varepsilon_j(s)}{2^j}$$

be the dyadic expansion of s (i.e. $\varepsilon_0(t) \in I\!N^0, \varepsilon_j(t) \in \{0, 1\}$ $(j \ge 1)$, and $\varepsilon_j(t) = 0$ for arbitrary large j). Write

$$s_r = \frac{\lfloor 2^r s \rfloor}{2^r} = \sum_{j=0}^{r} \frac{\varepsilon_j(s)}{2^j},$$

$$R = 2^r.$$

For fixed s, t, r, we have a.s.

$$\begin{aligned}
2\sqrt{|q((-B(s)) \cdot B(s+t))|} \le\ & |(-B(s_r)) \cdot B((s+t)_r)| + |(-B((s+t)_r)) \cdot B(s+t)| \\
& + |(-B(s)) \cdot B(s_r)| \\
\le\ & |(-B(s_r)) \cdot B((s+t)_r)| \\
& + \sum_{j=0}^{\infty} |(-B((s+t)_{r+j})) \cdot B((s+t)_{r+j+1})| \\
& + \sum_{j=0}^{\infty} |(-B(s_{r+j})) \cdot B(s_{r+j+1})|
\end{aligned} \tag{2.54}$$

Since

$$\sup_{0<t<h} |(s+t)_r - s_r| \le h + \frac{1}{R},$$

$$\sup_{0<t<h} |(s+t)_{r+j+1} - (s+t)_{r+j}| \le \frac{1}{2^{r+j+1}},$$

there are, by the form of the density of $A(1)$, $Q, C_1 > 0$ such that for any $h, u, x_j > 0$ and $r, j \in \mathbb{N}^0$

$$\begin{aligned} &P(\sup_{0<s<1-h}\sup_{0<t<h}|(-B(s_r))\cdot B((s+t)_r)| \geq u\sqrt{h+\frac{1}{R}}) \\ &\leq 2\frac{R}{Rh+1}P(|B(h+\frac{1}{R})| \geq u\sqrt{h+\frac{1}{R}}) \\ &= 2\frac{R}{Rh+1}P(|B(1)| \geq u) \\ &\leq C_1\frac{R}{Rh+1}e^{-Qu^2} \end{aligned}$$

and analogously

$$P(\sup_{0\leq s\leq 1-h}\sup_{0<t<h}|(-B((s+t)_{r+j}))\cdot B((s+t)_{r+j+1})| \geq \frac{x_j}{\sqrt{2^{r+j+1}}}) \leq C_1\cdot 2^{r+j+1}e^{-Qx_j^2},$$

$$P(\sup_{0\leq s\leq 1-h}\sup_{0<t\leq h}|(-B(s_{r+j}))\cdot B(s_{r+j+1})| \geq \frac{x_j}{\sqrt{2^{r+j+1}}}) \leq C_1\cdot 2^{r+j+1}e^{-Qx_j^2}.$$

Thus by (2.54)

$$\begin{aligned} &P(\sup_{0<s<1-h}\sup_{0<t<h} 2\sqrt{|q((-B(s))\cdot B(s+t))|} \geq u\sqrt{1+\frac{1}{R}}+2\sum_{j=0}^{\infty}\frac{x_j}{\sqrt{2^{r+j+1}}}) \\ &\leq C_1\frac{R}{Rh+1}e^{-Qu^2}+4C_1R\sum_{j=0}^{\infty}2^je^{-Qx_j^2}. \end{aligned} \tag{2.55}$$

Now put

$$x_j = \sqrt{\frac{j}{Q}+u^2}$$

and R such that

$$2R > \frac{K}{h} \geq R, \tag{2.56}$$

where K is a large enough constant. Then we get

$$4C_1R\sum_{j=0}^{\infty}2^je^{-Qx_j^2} \leq \frac{C_2K}{h}e^{-Qu^2}. \tag{2.57}$$

and (as in Csörgö, Révész (1981), p.25))

$$u\sqrt{h+\frac{1}{R}}+2\sum_{j=0}^{\infty}\frac{x_j}{\sqrt{2^{r+j+1}}} \leq \sqrt{h}\{u|\sqrt{1+\frac{2}{K}}+\frac{2G}{\sqrt{K}}|+\frac{2B}{\sqrt{K}}\} \tag{2.58}$$

holds. Now put

$$v = \frac{1}{4}\{u|\sqrt{1+\frac{2}{K}}+\frac{2G}{\sqrt{K}}|+\frac{2B}{\sqrt{K}}\}^2, \tag{2.59}$$

so from (2.55)-(2.59), for $v \geq 1$,

$$P(\sup_{0<s<1-h}\sup_{0<t\leq h}|q((-B(s))\cdot B(s+t))| \geq vh) \leq \frac{D}{h}e^{-Qu^2} \leq \frac{D}{h}\exp[-\frac{v}{H+\varepsilon}]$$

with $H := (4Q)^{-1}$ by taking K large enough.□

The following corollary is immediate by the fact that $\mathcal{L}(\{B(t)\}_{0\leq t\leq T}) = \mathcal{L}(\{TB(t/T)\}_{0\leq t\leq T}$ (cf. Csörgö, Révész (1981), Lemma 1.2.1, Neuenschwander, Schott (1995), Corollary 1):

Corollary 2.3 *There are constants $H > 0$ and, for every $\varepsilon > 0$, a $D = D(\varepsilon) > 0$ such that for every $v > 0$, $T > 0$, $0 < h < T$*

$$P(\sup_{0\leq s\leq T-h}\sup_{0\leq t\leq h}|q((-B(s))\cdot B(s+t))| \geq vh) \leq \frac{DT}{h}\exp[-\frac{v}{H+\varepsilon}].$$

Proof of Theorem 2.20: Again the proof follows closely the proof of Theorem 1.2.1 in Csörgö, Révész (1981) in an adapted simplified version, using the formula for the density of $A(1)$ and Corollary 2.3 instead of Csörgö, Révész (1981), (1.1.1) and Lemma 1.2.1.

1. We first prove the upper bound. Let

$$\alpha(T) := \sup_{0\leq t\leq T-c\log T}\sup_{0\leq s<c\log T}\frac{|q((-B(t))\cdot B(t+s))|}{(\log T)^2}.$$

Assume $\varepsilon > 0$. By Corollary 2.3 (with $h = c\log T, v = (H+2\varepsilon)\log T$) we get, for T big enough,

$$\begin{aligned}
&P(\alpha(T) \geq (H+2\varepsilon)c) \\
&= P(\sup_{0<t\leq T-c\log T}\sup_{0\leq s\leq c\log T}|q((-B(t))\cdot B(t+s))| \geq (H+2\varepsilon)c(\log T)^2) \\
&< \frac{DT}{c\log T}\exp[-\frac{H+2\varepsilon}{H+\varepsilon}\log T] \\
&\leq \frac{DT}{c\log T}\exp[-\frac{H+2\varepsilon}{H+\varepsilon}\log(\frac{T}{c\log T})] \\
&= D(\frac{T}{c\log T})^{-\varepsilon/(H+\varepsilon)}.
\end{aligned}$$

Putting $T_k := \theta^k$ $(\theta > 1)$ yields, by the Borel-Cantelli lemma,

$$\limsup_{k\to\infty}\alpha(T_k) \leq Hc.$$

Since $T \mapsto (\log T)^2$ and $T \mapsto (\log T)^2\alpha(T)$ are (a.s.) non-decreasing and since $1 \leq (\log T_{k+1})/\log T_k \leq \theta$ (k big enough), choosing θ near enough to 1 yields

$$\limsup_{T\to\infty}\alpha(T) \leq Hc.$$

2. Now we show the lower bound. Put, for $S, T > 0$

$$\beta(S,T) := \frac{|q((-B(S - c\log T)) \cdot B(S))|}{c(\log T)^2}.$$

By the estimation

$$P(A(1) > u) \geq \frac{1}{2\pi} e^{-\pi u} \tag{2.60}$$

(cf. Helmes (1986), (21)) we get, for $\varepsilon > 0$ and T big enough,

$$\begin{aligned} P(\beta(S,T) \geq \frac{1-2\varepsilon}{\pi}) &\geq \frac{1}{\pi}\exp[-c(\log T)^2(1-2\varepsilon)/(c\log T)] \\ &= \frac{1}{\pi} T^{2\varepsilon - 1} \\ &\geq \frac{1}{\pi}(\frac{c\log T}{T})^{1-\varepsilon}, \end{aligned}$$

hence

$$\begin{aligned} P(\max_{0<k\leq\lfloor T/(c\log T)\rfloor - 1} \frac{|q((-B(kc\log T)) \cdot B((k+1)c\log T))|}{c(\log T)^2} \leq \frac{1-\varepsilon}{\pi}) \\ \leq (1 - \frac{1}{\pi}(\frac{c\log T}{T})^{1-\varepsilon})^{T/(c\log T)} \\ \leq 2\exp[-\frac{1}{\pi}(\frac{T}{c\log T})^{\varepsilon}]. \end{aligned}$$

Since

$$\sum_{j=2}^{\infty} \exp[-\frac{1}{\pi}(\frac{j}{c\log j})^{\varepsilon}] < \infty,$$

we get

$$\begin{aligned} \liminf_{j\to\infty} \alpha(j) &\geq \liminf_{j\to\infty} \max_{0\leq k\leq\lfloor j/(c\log j)\rfloor - 1} \frac{|q((-B(kc\log j)) \cdot B((k+1)c\log j))|}{(\log j)^2} \\ &\geq \frac{c}{\pi}. \end{aligned}$$

Now for the in-between-times $j \leq T \leq j+1$, we may do the analogous calculations as in Csörgö, Révész (1981), p.34 using 1. to get

$$\liminf_{T\to\infty} \alpha(T) \geq \frac{c}{\pi}. \square$$

Now we will give a qualitative analogue of Theorem 2.18 (iii) (cf. Neuenschwander, Schott (1995), Theorem 3).

Theorem 2.21

$$\inf_{0\leq s\leq 1-h} \sup_{0\leq t\leq h} \frac{\log(1/h)}{h} |q((-B(s)) \cdot B(s+t))| = \Theta(1) \quad (h \to 0).$$

Proof: The proof is similar as in Csörgö, Révész (1981), pp.45ff. mutatis mutandis using Lemma 2.8 and the form of the density of $A(1)$ together with (2.60) instead of Csörgö, Révész (1981), Lemma 1.6.1. $\square$

We conjecture that Theorem 2.20 (and then also Theorem 2.21 (in contrast to the real-valued situation!)) hold with "$\lim \ldots = 1/\pi$" (that is what one would obtain if $H = 1/\pi$ in Lemma 2.12).

Let $h : [0,1] \to [0,\infty]$ be a continuous increasing function with $h(0) = 0$ and assume $E \subset I\!R^d$. Then the *Hausdorff h-measure* of E is given by

$$h - m(E) := \lim_{\delta > 0} \inf_{\substack{E \subset \bigcup_{i=1}^{\infty} C_i \\ d(C_i) < \delta}} \sum_{i=1}^{\infty} h(d(C_i)),$$

where $d(C_i)$ denotes the diameter of $C_i \subset I\!R^d$. Put, for $r > 0$,

$$h_r(t) := \begin{cases} t^2(\log \frac{1}{t} \vee 1)^r & : \quad t > 0, \\ 0 & : \quad t = 0. \end{cases}$$

Let $R \subset I\!H \cong I\!R^3$ be a bounded domain in $I\!R^3$ and

$$\tau := \inf\{t \geq 0 : B(t) \notin R\}.$$

The following theorem is Theorem 8.1 of Chaleyat-Maurel, Le Gall (1989):

Theorem 2.22 *For every $x_0 \in R, t > 0, t > 1$ we have*

$$h_1 - m(\{x_0 \cdot B(s) : 0 \leq s < t \wedge \tau\}) \overset{a.s.}{<} \infty. \tag{2.61}$$

and, for $r > 1$,

$$h_r - m(\{x_0 \cdot B(s) : 0 \leq s < t \wedge \tau\}) \overset{a.s.}{=} \infty. \tag{2.62}$$

Proof (cf. Chaleyat-Maurel, Le Gall (1989)): Notation is as in 1.3 and 2.2.1. Let $J_t := \{x_0 \cdot B(s) : 0 \leq s < t \wedge \tau\}$. We first prove (2.61). For $n \geq 1$, let N_n be the number of dyadic cubes of volume 2^{-n} which are contained in R and have a non-void intersection with J_t. Then

$$2^{-3n} N_n \leq \lambda(S_{U(0,3 \cdot 2^{-n})}(0,t)),$$

where $U(z,r) := \{x \in I\!R^3 : d(x,z) \leq r\}$. Furthermore by the definition of Hausdorff measures

$$\begin{aligned} h_1 - m(J_t) &\leq \liminf_{n \to \infty} h_1(3 \cdot 2^{-n}) N_n \\ &\leq 9 \log 2 \liminf_{n \to \infty} n 2^{-2n} N_n, \end{aligned}$$

hence by Theorem 2.8

$$\begin{aligned} h_1 - m(J_t) &\leq 9 \log 2 \liminf_{n \to \infty} n 2^{n} \lambda(S_{U(0,3 \cdot 2^{-n})}(0,t)) \\ &< \infty. \end{aligned}$$

Now we prove (2.62). Let H be a compact subset of R containing x_0 in its interior. Let ν be the occupation measure of $\{x_0 \cdot B(s) : 0 \leq s < t \wedge \tau\}$ restricted to H, i.e.

$$\nu(M) := \int_0^{t\wedge\tau} 1\{x_0 \cdot B(s) \in H \cap M\} ds$$

(M a Borel subset of $\mathbb{R}^3$). For $p \in \mathbb{N}$ there is a constant $C_p > 0$ such that for all $a \in]0, \frac{1}{2}[$ we have

$$E_{x_0}(\int_H \nu(U(y,a))^p \nu(dy)) \leq C_p a^{2p} (\log \frac{1}{a})^p. \tag{2.63}$$

Indeed: By the Markov property and induction on p we get

$$
\begin{aligned}
& E_{x_0}(\int_H \nu(U(y,a))^p \nu(dy)) \\
= \; & E_{x_0}(\int_0^t 1\{x_0 \cdot B(s) \in H\} \\
& \cdot (\int_0^t 1\{x_0 \cdot B(u) \in U(x_0 \cdot B(s), a) \cap H\} du)^p ds) \\
\leq \; & (p+1)! E_{x_0}(\int_{0<s_1<s_2<\ldots<s_{p+1}<t} \prod_{i=1}^{p} 1\{x_0 \cdot B(s_i) \in H\} \\
& \cdot 1\{x_0 \cdot B(s_{i+1}) \in U(x_0 \cdot B(s_i), 2a)\} ds_1\, ds_2 \ldots ds_{p+1}) \\
\leq \; & (p+1)! E_{x_0}(\int_{0<s_1<s_2<\ldots<s_p<t} \prod_{i=1}^{p-1} 1\{x_0 \cdot B(s_i) \in H\} \\
& \cdot 1\{x_0 \cdot B(s_{i+1}) \in U(x_0 \cdot B(s_i), 2a)\} \\
& \cdot \int_{U(x_0 \cdot B(s_p), 2a)} \int_0^\infty p_t((-(x_0 \cdot B(s_p))) \cdot z) dt\, dz\, ds_1\, ds_2 \ldots ds_p) \\
\leq \; & (p+1)! t \sup_{y \in H}(\int_{U(y,2a)} \int_0^\infty p_t((-y) \cdot z) dt\, dz)^p.
\end{aligned}
$$

Now the local study of the distance d and estimates of the Green function $\int_0^\infty p_t((-y) \cdot z)dt$ (cf. Chaleyat-Maurel, Le Gall (1989), p. 260) yield a constant $C > 0$ such that for every $a \in]0, \frac{1}{2}[$ we have

$$\sup_{y \in H}(\int_{U(y,2a)} \int_0^\infty p_t((-y) \cdot z) dt\, dz) \leq C a^2 \log \frac{1}{a}.$$

Now assume $\{a_n\}_{n \geq 1} \in]0, \infty[$ is a sequence tending to 0 fast enough and let $r > 1$. Since the series

$$\sum_{n=1}^{\infty} (\log \frac{1}{a_n})^{p(1-r)}$$

is convergent, it follows from the Borel-Cantelli lemma that

$$\limsup_{a \to 0} \frac{\nu(U(y,a))}{h_r(a)} = 0 \quad P_{x_0} - a.s., \nu(dy) - a.e.$$

Now by density theorems for Hausdorff measures (cf. Chaleyat-Maurel, Le Gall (1989), p. 261 and Rogers, Taylor (1961))

$$h_r - m(J_t) \geq \nu(J_t) > 0 \quad P_{x_0} - a.s. \square$$

Theorem 2.22 is the same result as for planar Brownian motion on $\mathbb{R}^2$ (cf. Chaleyat-Maurel, Le Gall (1989), p.261). However, the situation is quite different if we consider the question of the existence of multiple points. The following theorem is Theorem 8.2 of Chaleyat-Maurel, Le Gall (1989), whose proof is similar to the corresponding proof for Brownian motion on $\mathbb{R}^4$ (cf. Chaleyat-Maurel, Le Gall (1989), Remark (i) on p. 263:

Theorem 2.23 *$\{B(t)\}_{t>0}$ has a.s. no double points.*

2.3.2 The Crépel-Roynette law of the iterated logarithm

Crépel and Roynette (1977) gave an analogue of the classical Hartman-Wintner law of the iterated logarithm for $\mathbb{H}$-valued random variables (X', X'', X''') such that X''' has a moment of order 1 and (X', X'') an absolute moment of order $2 + \rho \quad (\rho > 0)$. The idea of Crépel, Roynette (1977) was to prove an estimation of the speed of convergence in the central limit theorem on $\mathbb{H}$ based on such a theorem due to Sazonov for random variables in a euclidean space together with a "maximal lemma" similar to Lemma 2.8.

Theorem 2.24 *Let $X_1, X_2, \ldots$ be i.i.d. $\mathbb{H}$-valued random variables on some probability space $(\Omega, \mathcal{F}, P)$, $E(X_1) = m$, $E\|(X_1', X_1'')\|^{2+\rho} < \infty$ for some $\rho > 0$, $C := (E((X_1' - m')^2)E((X_1'' - m'')^2) - E^2((X_1' - m') \cdot (X_1'' - m'')))^{1/2} \in]0, \infty[$. Then the set of accumulation points of*

$$\{\pi q(\prod_{j=1}^{n} (X_j \cdot (-m)))/(n \log\log n)\}_{n \geq 1}$$

is a.s. equal to the interval $[-\frac{1}{C}, \frac{1}{C}]$.

First of all, one has to prove a theorem concerning the speed of convergence in the central limit theorem on $\mathbb{H}$, which is of independent interest.

Theorem 2.25 *There exists $c > 0$ such that*

$$|P(a \leq \frac{1}{n} q(\prod_{j=1}^{n} X_j \cdot (-m)) \leq b) - \int_a^b \frac{dx}{\cosh \pi x}| \leq c n^{-\frac{1}{3 \cdot 63/\rho}}$$

$(a, b \in [-\infty, \infty], n \geq 1)$.

Since $E(X_1 \cdot b) = E(X_1) \cdot b = 0$ we may w.l.o.g. assume $E(X_1) = 0$. The case $C = \infty$ is trivial. For every $v, w, \theta \in \mathbb{R}$, $vw \neq 0$, the map $\mathbb{H} \to \mathbb{H}$ given by the matrix

$$\begin{pmatrix} v\cos\theta & v\sin\theta & 0 \\ -w\sin\theta & w\cos\theta & 0 \\ 0 & 0 & vw \end{pmatrix}$$

is an isomorphism of $\mathbb{H}$. By this and the strong law of large numbers (applied to $X_1''', X_2''', \ldots$) we may furthermore w.l.o.g. assume that $E(X_1'^2) = E(X_1''^2) = 1$, $E(X_1'X_1'') = 0, X_n''' \overset{a.s.}{=} 0 \quad (n \geq 1)$ and thus $C = 1$.
Put

$$N_p^{(k)} := \sum_{\substack{i,j=1 \\ i<j}}^{k} [X_{pk+i}, X_{pk+j}],$$

$$Q_p^{(k)} := \sum_{j=1}^{(p+1)k} X_j.$$

With this we get

$$q(\prod_{j=1}^{nk} X_j) = \frac{1}{2} q(\sum_{p=0}^{n-1} N_p^{(k)}) + \frac{1}{2} q(\sum_{p=0}^{n-2} [Q_p^{(k)}, Q_{p+1}^{(k)}]). \tag{2.64}$$

Lemma 2.13 *Put*

$$R_n := (X_1', X_1' + X_2', \ldots, \sum_{j=1}^{n} X_j', X_1'', X_1'' + X_2'', \ldots, \sum_{j=1}^{n} X_j'').$$

Then

$$E(||R_n||^{2+\rho}) \leq c_1 n^6$$

for some constant c_1.

Proof: If one expands $E||R_n||^2$, one gets $p_n = 2\sum_{k=1}^n k^2 - \frac{n(n-1)(2n-1)}{3} \leq n^3$ terms of the form $E(X_k'X_\ell')$ resp. $E(X_k''X_\ell'')$. Observe that

$$\begin{aligned} |\sum_{k=1}^{p} a_k|^{1+\rho/2} &= |\sum_{k=1}^{p} a_k| |\sum_{k=1}^{p} a_k|^{\rho/2} \\ &\leq \sum_{k=1}^{p} |a_k| \sum_{k=1}^{p} |a_k|^{\rho/2} \\ &\leq p \sum_{k=1}^{p} |a_k|^{1+\rho/2}. \end{aligned} \tag{2.65}$$

Since $|xy|^{1+\rho/2} \leq \frac{1}{2}(|x|^{2+\rho} + |y|^{2+\rho}) \quad (x, y \in \mathbb{R})$, we thus get, by substituting $E(X_k'X_\ell')$ and $E(X_k''X_\ell'')$ for the a_k in (2.65)

$$\begin{aligned} E(||R_n||^{2+\rho}) &\leq \frac{1}{2} p_n^2 (E|X_1'|^{2+\rho} + E|X_1''|^{2+\rho}) \\ &\leq c_1 n^6. \square \end{aligned}$$

Lemma 2.14 *Let* $\overline{Q_0}, \overline{Q_1}, \ldots$ *be* $\mathbb{R}^2$*-valued random variables such that* $(\overline{Q_0}', \overline{Q_0}'', \overline{Q_1}', \overline{Q_1}'', \ldots, \overline{Q_{n-1}}', \overline{Q_{n-1}}'')$ $(x = (x', x'') \in \mathbb{R}^2)$ *is centered Gaussian with covariance matrix*

$$\Delta_n = \begin{pmatrix} \Omega_n & 0 \\ 0 & \Omega_n \end{pmatrix},$$

$$\Omega_n = \begin{pmatrix} 1 & 1 & \ldots & 1 \\ 1 & 2 & \ldots & 2 \\ \ldots & \ldots & \ldots & \ldots \\ 1 & 2 & \ldots & n \end{pmatrix}.$$

Then there exists a constant C_2 *such that*

$$|P(a \le \frac{1}{nk} q(\sum_{p=0}^{n-2} [Q_p^{(k)}, Q_{p+1}^{(k)}]) \le b) - P(a \le \frac{1}{n} q(\sum_{p=0}^{n-2} [(\overline{Q_p}, 0), (\overline{Q_{p+1}}, 0)]) \le b) \le c_2 \frac{n^{10}}{k^{\rho/2}}$$

$(a, b \in [-\infty, \infty], n, k \ge 1)$.

Proof: By a theorem due to Sazonov (cf. Crépel, Roynette (1977), p.220) and Lemma 2.13, we get

$$\begin{aligned}
&|P(\frac{1}{\sqrt{k}}(Q_0^{(k)'}, Q_1^{(k)'}, \ldots, Q_{n-1}^{(k)'}, Q_0^{(k)''}, Q_1^{(k)''}, \ldots, Q_{n-1}^{(k)''}) \in H) \\
&\quad - P((\overline{Q_0}', \overline{Q_1}', \ldots, \overline{Q_{n-1}}', \overline{Q_0}'', \overline{Q_1}'', \ldots, \overline{Q_{n-1}}'') \in H) \\
&\le \frac{c_3 n^4}{k^{\rho/2}} E\|R_n\|^{2+\rho} \\
&\le \frac{c_3 n^4}{k^{\rho/2}} c_1 n^6 \\
&\quad c_2 \frac{n^{10}}{k^{\rho/2}}
\end{aligned}$$

for every subset $H \subset \mathbb{R}^{2n}$ whose boundary is a hyperquadric, $n, k \ge 1$. Now the assertion follows from the fact that the set

$$H_r = \{(x_0', x_1', \ldots, x_{n-1}', x_0'', x_1'', \ldots, x_{n-1}'') \in \mathbb{R}^{2n} : q(\sum_{p=0}^{n-2} [(x_p, 0), (x_{p+1}, 0)]) = r\}$$

$(x = (x', x'') \in \mathbb{R}^2)$ is a hyperquadric. □

Lemma 2.15 *In the situation of Lemma 2.14 there exists a constant* $c_4 > 0$ *such that*

$$|P(a \le \frac{1}{n} q(\sum_{p=0}^{n-2} [(\overline{Q_p}, 0), (\overline{Q_{p+1}}, 0)]) \le b) - \int_a^b \frac{dx}{\cosh \pi x}| \le \frac{c_4}{n^{1/3}} \tag{2.66}$$

for all $a, b \in [-\infty, \infty], n \ge 1$.

Proof: Let

$$A := \begin{pmatrix} 0 & 1 & \dots & 1 & 1 \\ -1 & 0 & \dots & 0 & 1 \\ \dots & \dots & \dots & \dots & \dots \\ -1 & 0 & \dots & 0 & 1 \\ -1 & -1 & \dots & -1 & 0 \end{pmatrix},$$

$$M := \begin{pmatrix} 0 & A \\ A^{tr} & 0 \end{pmatrix}.$$

Then one sees that

$$\mathcal{L}(\frac{1}{n}q(\sum_{p=0}^{n-2}|(\overline{Q_p},0),(\overline{Q_{p+1}},0)|)) = \mathcal{L}(\frac{1}{2n}X^{tr}MX),$$

where X is an $\mathbb{R}^{2n}$-valued random vector obeying to a standard Gaussian law. It is known that the Fourier transform of $\mathcal{L}(\frac{1}{2n}X^{tr}MX)$ is given by

$$\begin{aligned} \varphi_n(u) &= |\det(I - \frac{iu}{2n}M)|^{-1/2} \\ &= |\det(I - \frac{u^2}{4n^2}A^2)|^{-1/2} \\ &= |\det(I - \frac{u}{2n}A)|^{-1/2} \cdot |\det(I + \frac{u}{2n}A)|^{-1/2} \\ &= |\det(I + \frac{u}{2n}A)|^{-1} \\ &= \{\frac{1}{2}[(1+\frac{u}{2n})^n + (1-\frac{u}{2n})^n]\}^{-1}, \end{aligned}$$

the last equality being proved by recurrence. This implies by an elementary calculation

$$|\varphi_n(u) - (\cosh\frac{u}{2})^{-1}| \le \frac{u^2}{4n} \quad (u \le n),$$

which, together with a classical inequality comparing the difference of probabilities with the difference of their Fourier transforms (cf. Crépel, Roynette (1977), p.222), and the fact that the Fourier transform of the measure with density $(\cosh \pi x)^{-1}$ $(x \in \mathbb{R})$ is given by $(\cosh\frac{u}{2})^{-1}$ $(u \in \mathbb{R})$, implies that the left hand side of (2.66) is majorized by

$$\frac{2}{\pi}\int_0^T \frac{u}{4n}du + \frac{24}{\pi T} = \frac{T^2}{4\pi n} + \frac{24}{\pi T}$$

for every $T > 0$. Now the assertion follows by taking $T = n^{1/3}$. □
Lemmas 2.14 and 2.15 imply:

Lemma 2.16 *There exists $c_5 > 0$ such that*

$$|P(a \le \frac{1}{nk}q(\sum_{p=0}^{n-2}|Q_p^{(k)},Q_{p+1}^{(k)}|) \le b) - \int_a^b \frac{dx}{\cosh \pi x}| \le \frac{c_5}{n^{1/3}}$$

for all $a, b \in [-\infty, \infty], n \ge 1, k \ge n^{21/\rho}$.

Now we are going to prove

Lemma 2.17 *There exists $c_6 > 0$ such that*

$$|P(a \le \frac{1}{nk} q(\prod_{j=1}^{nk} X_j) \le b) - \int_a^b \frac{dx}{\cosh \pi x}| \le \frac{c_6}{n^{1/3}}$$

for all $a, b \in [-\infty, \infty], n \ge 1, k \ge n^{21/\rho}$.

Proof: By (2.64)

$$\begin{aligned}
&|P(a \le \frac{1}{nk} q(\prod_{j=1}^{nk} X_j) \le b) - P(a \le \frac{1}{nk} q(\sum_{p=0}^{n-2} [Q_p^{(k)}, Q_{p+1}^{(k)}]) \le b)| \\
&= |\Big(\int\limits_{\frac{1}{2nk} q(\sum_{p=0}^{n-1} N_p^{(k)}) > \varepsilon} + \int\limits_{\frac{1}{2nk} q(\sum_{p=0}^{n-1} N_p^{(k)}) < \varepsilon} \Big) \Big(1\{a \le \frac{1}{nk} q(\prod_{j=1}^{nk} X_j) \le b\} \\
&\quad - 1\{a \le \frac{1}{nk} q(\sum_{p=0}^{n-2} [Q_p^{(k)}, Q_{p+1}^{(k)}]) \le b\}\Big) P(d\omega)| \\
&\le P(q(\sum_{p=0}^{n-1} N_p^{(k)}) > 2\varepsilon nk) \\
&\quad + P(\frac{1}{nk} q(\sum_{p=0}^{n-2} [Q_p^{(k)}, Q_{p+1}^{(k)}]) \in [a-\varepsilon, a+\varepsilon] \cup [b-\varepsilon, b+\varepsilon]). \qquad (2.67)
\end{aligned}$$

Since

$$E(q(\sum_{p=0}^{n-1} N_p^{(k)})^2) = O(nk^2) \quad (n \to \infty),$$

the first summand on the right hand side of (2.67) is majorized by $c_7/(n\varepsilon^2)$ by Čebyšev's inequality. By Lemma 2.16, the second one may be estimated from above by

$$2\frac{c_5}{n^{1/3}} + \Big(\int_{a-\varepsilon}^{a+\varepsilon} + \int_{b-\varepsilon}^{b+\varepsilon}\Big) \frac{dx}{\cosh \pi x} \le 2\frac{c_5}{n^{1/3}} + 4\varepsilon.$$

So, by taking $\varepsilon = n^{-1/3}$, the term on the right hand side of (2.67) is not larger than

$$\frac{c_7}{n\varepsilon^2} + 2\frac{c_5}{n^{1/3}} + 4\varepsilon \le \frac{c_7}{n^{1/3}} + \frac{2c_5}{n^{1/3}} + \frac{4}{n^{1/3}} = \frac{c_6}{n^{1/3}}. \square$$

Proof of Theorem 2.25: Let $k := \lceil n^{21/\rho} \rceil, \varepsilon := k^{1/3}$ and decompose $p = nk + r, 0 \le r < n$. Then we get

$$q(\prod_{j=1}^{p} X_j) = q(\prod_{j=1}^{nk} X_j) + q(V_n^{(k)}),$$

$$V_n^{(k)} = \frac{1}{2} \sum_{j=1}^{r} \sum_{i=1}^{nk+j-1} [X_i, X_{nk+j}].$$

and thus

$$
\begin{aligned}
& P(a \leq \frac{1}{p} q(\prod_{j=1}^{p} X_j) \leq b) - P(a \leq \frac{1}{nk} q(\prod_{j=1}^{nk} X_j) \leq b) \\
& - \; P(a(1+\frac{r}{nk}) \leq \frac{1}{nk} q(\prod_{j=1}^{nk} X_j) + \frac{1}{nk} q(V_n^{(k)}) \leq b(1+\frac{r}{nk})) \\
& \quad - P(a \leq \frac{1}{nk} q(\prod_{j=1}^{nk} X_j) \leq b) \\
& = \; (\int\limits_{\frac{1}{nk}|q(V_n^{(k)})|>\varepsilon} + \int\limits_{\frac{1}{nk}|q(V_n^{(k)})| \leq \varepsilon})(1\{a(1+\frac{r}{nk}) \\
& \quad \leq \frac{1}{nk} q(\prod_{j=1}^{nk} X_j) + \frac{1}{nk} q(V_n^{(k)}) \leq b(1+\frac{r}{nk})\} \\
& \quad -1\{a \leq \frac{1}{nk} q(\prod_{j=1}^{nk} X_j) \leq b\})P(d\omega). \qquad (2.68)
\end{aligned}
$$

By Čebyšev's inequality and a simple calculation, one sees that the absolute value of the first integral on the right hand side of (2.68) is majorized by

$$
\begin{aligned}
P(|q(V_n^{(k)})| > \varepsilon nk) & \leq \frac{E(q(V_n^{(k)})^2)}{\varepsilon^2 n^2 k^2} \\
& \leq \frac{c_8}{\varepsilon^2 k}, \qquad (2.69)
\end{aligned}
$$

whereas by Lemma 2.17 the absolute value of the second one is not larger than

$$
\begin{aligned}
& \frac{2c_6}{n^{1/3}} + (\int_{a-\varepsilon}^{a-\varepsilon+\frac{|a|r}{nk}} + \int_{b-\varepsilon}^{b+\varepsilon-\frac{|b|r}{nk}}) \frac{dx}{\cosh \pi x} \\
\leq \; & \frac{2c_6}{n^{1/3}} + (\int_{a-\varepsilon}^{a+\varepsilon+\frac{|a|}{k}} + \int_{b-\varepsilon}^{b-\varepsilon+\frac{|b|}{k}}) \frac{dx}{\cosh \pi x} \\
\leq \; & \frac{2c_6}{n^{1/3}} + 2 \sup\{3\varepsilon, \frac{2}{\pi} e^{-\pi(k-1)\varepsilon}\} \qquad (2.70)
\end{aligned}
$$

(consider the case $\frac{|a|}{k} \leq \varepsilon$ and $\frac{|a|}{k} > \varepsilon$ separately). Putting Lemma 2.17 and (2.68)-(2.70) together, we thus obtain

$$
|P(a \leq \frac{1}{p} q(\prod_{j=1}^{p} X_j) \leq b) - \int_a^b \frac{dx}{\cosh \pi x}| \leq \frac{c_9}{n^{1/3}} \leq c p^{\frac{1}{3+63/\rho}}
$$

$(a, b \in [-\infty, \infty], p \in \mathbb{N})$.□

In order to continue, we need the following maximal lemma (which is similar to Lemma 2.8 and Schott (1981), Lemma 3):

Lemma 2.18 *There are constants $b, c > 0$ such that for every $a \in \mathbb{R}$ and every $n \geq 1$*

$$P(\sup_{1 \leq k \leq n} q(\prod_{j=1}^{k} X_j) > a) \leq \frac{1}{c} P(q(\prod_{j=1}^{n} X_j) > a - bn).$$

Proof: For $1 \leq k \leq n$ consider the events

$$A_k := \{q(X_1) \leq a, q(X_1 \cdot X_2) \leq a, \dots, q(\prod_{j=1}^{k-1} X_j) \leq a, q(\prod_{j=1}^{k} X_j) > a\}.$$

Then

$$\begin{aligned} P(q(\prod_{j=1}^{n} X_j) > a - bn) &\geq \sum_{k=1}^{n} P((q(\prod_{j=1}^{n} X_j) > a - bn) \cap A_k) \\ &\geq \sum_{k=1}^{n} P((q(\prod_{j=1}^{n} X_j - \prod_{j=1}^{k} X_j) > -bn) \cap A_k). \end{aligned}$$

Hence, as

$$P(\sup_{1 < k < n} q(\prod_{j=1}^{k} X_j) > a) = \sum_{k=1}^{n} P(A_k),$$

it suffices to prove that for all $n \geq 1, 1 \leq k \leq n$

$$P((q(\prod_{j=1}^{n} X_j - \prod_{j=1}^{k} X_j) > -bn) \cap A_k) \geq cP(A_k). \tag{2.71}$$

One gets

$$\prod_{j=1}^{n} X_j - \prod_{j=1}^{k} X_j = \prod_{j=k+1}^{n} X_j + \frac{1}{2}[\sum_{j=1}^{k} X_j, \sum_{j=k+1}^{n} X_j].$$

If we condition on $\{X_1, X_2, \dots, X_k\}$, then (2.71) is proved by showing that for all $x \in \mathbb{H}^d, n \geq 1, 1 \leq k \leq n$

$$P(q(\prod_{j=k+1}^{n} X_j + [x, \sum_{j=k+1}^{n} X_j]) > -bn) \geq c.$$

The probability of the complementary event is majorized by

$$P(q(\prod_{j=k+1}^{n} X_j) \leq -\frac{b}{2}n) + P(q([x, \sum_{j=k+1}^{n} X_j]) \leq -\frac{b}{2}n) \tag{2.72}$$

It is easy to see that we may assume that $n - k$ is big enough. Now the assertion follows, for b big enough, from (2.72) and the central limit theorem (for both $\mathbb{R}$ and $\mathbb{H}$) (replace $-\frac{b}{2}n$ by $-\frac{b}{2}(n-k)$ in (2.72)). □

Now we continue the proof of Theorem 2.24:

End of proof of Theorem 2.24: 1. We first show

$$\limsup_{n \to \infty} \pi q(\prod_{j=1}^{n} X_j)/(n \log\log n) \overset{a.s.}{\leq} 1. \tag{2.73}$$

Let $\varepsilon > 0$. It suffices to show

$$P(\limsup_{n\to\infty}\{q(\prod_{j=1}^{n} X_j) > \frac{(1+\varepsilon)n\log\log n}{\pi}\}) = 0. \tag{2.74}$$

Let $d > 1$ be sufficiently near to 1. Then we get

$$\begin{aligned} & P(\limsup_{n\to\infty}\{q(\prod_{j=1}^{n} X_j) > \frac{(1+\varepsilon)n\log\log n}{\pi}\}) \\ \leq\ & P(\limsup_{k\to\infty}\{\sup_{0<p\leq d^k} q(\prod_{j=1}^{p} X_j) > \frac{1+\varepsilon}{\pi} d^{k-1}\log\log d^{k-1}\}) \\ \leq\ & P(\limsup_{k\to\infty}\{\sup_{0\leq p\leq d^k} q(\prod_{j=1}^{p} X_j) > \frac{\sigma}{\pi} d^{k}\log\log d^{k}\}) \end{aligned}$$

with $\sigma := (\frac{1+\varepsilon}{d} + 1)/2 > 1$. Now by Lemma 2.18, Theorem 2.25, and the estimation $(\cosh \pi x)^{-1} \leq 2e^{-\pi x}$, we have (for suitable $\sigma' > 1$)

$$\begin{aligned} & \sum_{k=0}^{\infty} P(\sup_{0\leq p<d^k} q(\prod_{j=1}^{p} X_j) > \frac{\sigma}{\pi} d^k \log\log d^k) \\ \leq\ & \frac{1}{c}\sum_{k=0}^{\infty} P(q(\prod_{j=1}^{d^k} X_j) > \frac{\sigma}{\pi} d^k\log\log d^k - bd^k) \\ \leq\ & \frac{1}{c}\sum_{k=0}^{\infty} P(q(\prod_{j=1}^{d^k} X_j) > \frac{\sigma'}{\pi} d^k\log\log d^k) \\ \leq\ & \frac{1}{c}\sum_{k=0}^{\infty}\Big(\int_{\frac{\sigma'}{\pi}\log\log d^k}^{\infty} \frac{dx}{\cosh \pi x} + cd^{-\frac{k+1}{3-63/p}}\Big) \\ \leq\ & H + \frac{K}{c}\sum_{k=0}^{\infty} k^{-\sigma'} \\ <\ & \infty, \end{aligned}$$

so the Borel-Cantelli lemma implies (2.74) and thus (2.73).
2. It remains to prove that every $\alpha \in [-1,1]$ is an accumulation point of $\{\pi q(\prod_{j=1}^{n} X_j)/(n\log\log n)\}_{n\geq 1}$. W.l.o.g. assume $\alpha > 0$. Pick $\varepsilon > 0$ and let d be big enough. Put

$$A_k := \{|\frac{\pi q(\prod_{j=1}^{d^k} X_j)}{d^k\log\log d^k} - \alpha| < \varepsilon\}.$$

We show

$$P(\limsup_{n\to\infty} A_k) = 1. \tag{2.75}$$

For this, it suffices to prove

$$P(\limsup_{n\to\infty} A'_k) = 1, \tag{2.76}$$

where

$$A_k' := \{ |\frac{\pi\, q(\prod_{j=1}^{d^k} X_j - \prod_{j=1}^{d^{k-1}} X_j)}{d^k \log\log d^k} - \alpha| < \frac{\varepsilon}{2}\}.$$

and that

$$|\frac{\pi\, q(\prod_{j=1}^{d^{k-1}} X_j)}{d^k \log\log d^k}| \overset{a.s.}{<} \frac{\varepsilon}{2} \qquad (k \geq k_0). \tag{2.77}$$

(2.77) follows immediately from 1. As far as (2.76) is concerned, write

$$\begin{aligned} q(\prod_{j=1}^{d^k} X_j - \prod_{j=1}^{d^{k-1}} X_j) &= \frac{1}{2} q(|\sum_{j=1}^{d^{k-1}} X_j, \sum_{j=1}^{d^k - d^{k-1}} X_{d^{k-1}+j}|) \\ &\quad + \frac{1}{2} q(\sum_{\substack{i,j=1 \\ i<j}}^{d^k - d^{k-1}} [X_{d^{k-1}+i}, X_{d^{k-1}+j}]) \\ &=: W_k + V_k. \end{aligned} \tag{2.78}$$

Now

$$\begin{aligned} \frac{\pi W_k}{d^k \log\log d^k} &= \pi \cdot a_k q(|(\sum_{j=1}^{d^{k-1}} X_j)/(2d^{k-1}\log\log d^{k-1})^{1/2}, \\ &\quad (\sum_{j=1}^{d^k-d^{k-1}} X_{d^{k-1}+j})/(2(d^k - d^{k-1})\log\log(d^k - d^{k-1}))^{1/2}|), \end{aligned}$$

where

$$a_k = \frac{\sqrt{d^{k-1}(d^k - d^{k-1})}}{d^k} \frac{\sqrt{\log\log d^{k-1} \cdot \log\log(d^k - d^{k-1})}}{\log\log d^k} \leq \frac{1}{\sqrt{d}},$$

hence by the classical law of the iterated logarithm we get

$$|\frac{\pi W_k}{d^k \log\log d^k}| \leq \frac{\pi}{\sqrt{d}} < \frac{\varepsilon}{4} \tag{2.79}$$

by taking d large enough. On the other hand, by Theorem 2.25 and the estimation $(\cosh \pi x)^{-1} \geq e^{-\pi x} \quad (x \geq 0)$

$$\begin{aligned} &\sum_{k=0}^{\infty} P(\frac{\pi V_k}{d^k \log\log d^k} \in [\alpha - \frac{\varepsilon}{4}, \alpha + \frac{\varepsilon}{4}]) \\ &= \sum_{k=0}^{\infty} P(\frac{V_k}{d^k - d^{k-1}} \in [\frac{\alpha - \varepsilon/4}{\pi(1-1/d)} \log\log d^k, \frac{\alpha + \varepsilon/4}{\pi(1-1/d)} \log\log d^k]) \\ &\geq \sum_{k=0}^{\infty} (\int_{\frac{\alpha-\varepsilon/4}{\pi(1-1/d)}\log\log d^k}^{\frac{\alpha+\varepsilon/4}{\pi(1-1/d)}\log\log d^k} \frac{dx}{\cosh \pi x} + c\, d^{-\frac{k}{3+63/\rho}}) \\ &\geq H + K \sum_{k=0}^{\infty} k^{-\frac{\alpha-\varepsilon/4}{(1-1/d)}} / \log k \\ &= \infty \end{aligned} \tag{2.80}$$

if d is big enough. As the V_k are independent, (2.78)-(2.80) and the Borel-Cantelli lemma yield (2.76). Finally (2.76) and (2.77) imply (2.75). □

2.3.3 The subsequence principle

The following principle, the general idea of which goes back to Chatterji (1974), has been formulated in Aldous (1977), p.64: Given any a.s. limit theorem for i.i.d. random variables, one can state an analogous limit theorem holding for all subsequences of some subsequence of an arbitrarily dependent tight sequence of random variables. The precise form is as follows:

Definition 2.1 *A limit statute is a measurable subset A of $M^1(\mathbb{R}^d) \times (\mathbb{R}^d)^\infty$ such that, if $\{X_n\}_{n\geq 1}$ is an i.i.d. sequence with $\mathcal{L}(X_1) = \mu \in M^1(\mathbb{R}^d)$ on some probability space $(\Omega, \mathcal{F}, P)$, then*

$$(\mu, (X_1(\omega), X_2(\omega), \ldots)) \in A \quad P-a.s.$$

and

$$(\mu, x) \in A, \sum_{j=1}^{\infty} ||\hat{x}_j - x_j|| < \infty \Longrightarrow (\mu, \hat{x}) \in A$$

$$(x = (x_1, x_2, \ldots), \hat{x} = (\hat{x}_1, \hat{x}_2, \ldots) \in (\mathbb{R}^d)^\infty). \tag{2.81}$$

(Cf. Aldous (1977), pp.62f. and p.68, end of section 3.) For a sequence $\{X_n\}_{n>1}$ of $\mathbb{R}^d$-valued random variables on $(\Omega, \mathcal{F}, P)$ such that $\{\mathcal{L}(X_n)\}_{n\geq 1}$ is uniformly tight, let $\mu = \mu(\omega) : \Omega \to M^1(\mathbb{R}^d)$ be the *associated random measure* constructed in Aldous (1977). For the applications, the explicit construction of μ is in fact really of no importance (cf. Aldous (1977), p. 61 bottom), so we will not give further details here. What will be used is only Aldous (1977), Lemma 2:

Lemma 2.19

$$E|\mu(\omega)|^p \leq \limsup_{n\to\infty} E|X_n|^p \quad (0 < p < \infty).$$

Proposition 2.4 *Let $\{X_n\}_{n>1}$ be any uniformly tight sequence of $\mathbb{R}^d$-valued random variables on some probability space $(\Omega, \mathcal{A}, P)$ with associated random measure μ and let A be a limit statute. Then there exists a subsequence $\{n'\} \subset \{n\}$ such that for any subsequence $\{n''\} \subset \{n'\}$*

$$(\mu(\omega), \{X_{n''}(\omega)\}_{n''}) \in A \quad P-a.s.$$

(Cf. Aldous (1977), Theorem 3 and p.68, end of section 3 for the generalization to $\mathbb{R}^d$.)
As otherwise the following explanations may seem somewhat obscure at first glance otherwise, we show by the classical example of Komlós's (1967) strong law of large numbers how Proposition 2.4 is applied (cf. Aldous (1977), pp. 63-64): Define the limit statute

$$A := \{(\mu, x) \in M^1(\mathbb{R}) \times \mathbb{R}^\infty : \lim_{n\to\infty} \frac{1}{n}\sum_{j=1}^{n} x_j = \int_{-\infty}^{\infty} \xi\mu(d\xi)\}$$

$$\cup\{(\mu, x) \in M^1(\mathbb{R}) \times \mathbb{R}^\infty : \int_{-\infty}^{\infty} |\xi|\mu(d\xi) = \infty\}.$$

Now assume $\{X_n\}_{n>1}$ is any sequence of $\mathbb{R}$-valued random variables such that $\sup_{n>1} E|X_n| < \infty$. Then clearly $\{\mathcal{L}(X_n)\}_{n>1}$ is uniformly tight. So by Proposition 2.4, there exists a subsequence $\{n'\} \subset \{n\}_{n>1}$ such that for every subsequence $\{n(m)\}_{m\geq 1} \subset \{n'\}$ we have a.s. $\int_{-\infty}^{\infty} |\xi| \mu(d\xi) - \infty$ or

$$\frac{1}{n} \sum_{j=1}^{m} X_{n(j)} \longrightarrow \int_{-\infty}^{\infty} \xi \mu(d\xi) \quad (m \to \infty). \tag{2.82}$$

But by Lemma 2.19

$$\int_{-\infty}^{\infty} |\xi| \mu(d\xi) \leq \limsup_{n \to \infty} E|X_n| < \infty.$$

so there exists a subsequence $\{n'\} \subset \{n\}_{n>1}$ such that for every subsequence $\{n(m)\}_{m>1} \subset \{n'\}$ (2.82) holds a.s.

Now we want to show how our limit theorems for i.i.d. $\mathbb{H}$-valued random variables may be carried over using Proposition 2.4 similarly as the classical ones in Aldous (1977). We use the following lemma:

Lemma 2.20 *For* $y, x_1, \hat{x}_1, x_2, \hat{x}_2, \ldots \in \mathbb{H}$, *we have*

$$\left\| \prod_{j=1}^{n} (\hat{x}_j \cdot y) - \prod_{j=1}^{n} (x_j \cdot y) \right\| \leq \left(\|y\| + \sum_{j=1}^{n} \|\hat{x}_j - x_j\| \right) \cdot \left(1 + \frac{1}{2} \sum_{j=1}^{n} (\|\hat{x}_j\| + \|x_j\|) + n\|y\| \right).$$

Proof:

$$\left\| \prod_{j=1}^{n} (\hat{x}_j \cdot y) - \prod_{j=1}^{n} (x_j \cdot y) \right\|$$

$$= \left\| \sum_{j=1}^{n} (\hat{x}_j \cdot y) + \frac{1}{2} \sum_{1 < i < j < n} [\hat{x}_i + y, \hat{x}_j + y] - \sum_{j=1}^{n} (x_j \cdot y) - \frac{1}{2} \sum_{1 < i < j < n} [x_i + y, x_j + y] \right\|$$

$$- \left\| \sum_{j=1}^{n} (\hat{x}_j \cdot y - x_j \cdot y) + \frac{1}{2} \sum_{1 \leq i < j \leq n} ([\hat{x}_i - x_i, \hat{x}_j + y] + [x_i + y, \hat{x}_j - x_j]) \right\|$$

$$\leq \sum_{j=1}^{n} \|\hat{x}_j - x_j\| + \frac{1}{2} \sum_{j=1}^{n} (\|\hat{x}_j\| + \|x_j\|) \|y\|$$

$$+ \frac{1}{2} \sum_{j=1}^{n} \|\hat{x}_j - x_j\| \cdot \left(\sum_{j=1}^{n} (\|\hat{x}_j\| + \|x_j\|) + 2n\|y\| \right)$$

$$\leq \left(\|y\| + \sum_{j=1}^{n} \|\hat{x}_j - x_j\| \right) \cdot \left(1 + \frac{1}{2} \sum_{j=1}^{n} (\|\hat{x}_j\| + \|x_j\|) + n\|y\| \right). \quad \square$$

Applying Proposition 2.4 to the law of the iterated logarithm of Crépel, Roynette (cf. Theorem 2.24), we get

Corollary 2.4 *Assume* $\{X_n\}_{n\geq 1}$ *is a uniformly tight sequence of* $\mathbb{H}$-*valued random variables such that* $\sup_{j\geq 1} E\|X_j\| < \infty$ *and* $\sup_{j>1} E\|(X_j', X_j'')\|^{2+\rho} < \infty$ *for some* $\rho >$

0. *Then there exists a* $\mathbb{H}$*-valued random variable* α, *a random variable* β *with values in* $[0,\infty[$ *and a subsequence* $\{n'\} \subset \{n\}$ *such that for every subsequence* $\{n(m)\}_{m\geq 1} \subset \{n'\}$ *the set of accumulation points of*

$$\{\pi q(\prod_{j=1}^{m}(X_{n(j)} \cdot \alpha))/(m \log\log m)\}_{m\geq 1}$$

is a.s. equal to the interval $[-\beta,\beta]$.

(Cf. Chatterji (1974), Gapoškin (1972), Aldous (1977), Corollary 5 for the classical case.)

Proof: For $\nu \in M^1(\mathbb{H})$ define the numbers

$$\nu_{|1|} := \int_{\mathbb{H}} ||\xi||\nu(d\xi) \in [0,\infty],$$

$$\nu_1 := \int_{\mathbb{H}} \xi\nu(d\xi) \in \mathbb{H},$$

$$\begin{aligned}\nu_2 := &\Big(\Big(\int_{\mathbb{H}}(\xi'-\nu_1')^2\nu(d\xi)\Big)\Big(\int_{\mathbb{H}}(\xi''-\nu_1'')^2\nu(d\xi)\Big)\\ &-\Big(\int_{\mathbb{H}}(\xi'-\nu_1')(\xi''-\nu_1'')\nu(d\xi)\Big)^2\Big)^{1/2} \in [0,\infty],\end{aligned}$$

and put

$$\alpha(\nu) := -\nu_1,$$

$$\beta(\nu) := \nu_2.$$

Let $A \subset M^1(\mathbb{H}) \times \mathbb{H}^\infty$ be given by

$$A := A_1 \cup A_2 \cup A_3,$$

$$\begin{aligned}A_1 := \{&(\nu,x) \in M^1(\mathbb{H}) \times \mathbb{H}^\infty :\\ &\mathrm{acc}(\{\pi q(\prod_{j=1}^{n}(x_j \cdot \alpha(\nu)))/(n\log\log n)\}_{n>1})\\ &\overset{a.s.}{=} [-\beta(\nu),\beta(\nu)],\ \lim_{n\to\infty}\frac{1}{n}\sum_{j=1}^{n}||x_j|| < \infty,\ \nu_{|1|} < \infty\}\end{aligned}$$

(acc(...) denoting the set of accumulation points),

$$A_2 := \{(\nu,x) \in M^1(\mathbb{H}) \times \mathbb{H}^\infty : \int_{\mathbb{H}}((\xi'-\nu_1')^2 + (\xi''-\nu_1'')^2)^{(2+\rho)/2}\nu(d\xi) = \infty,\ \nu_{|1|} < \infty\},$$

$$A_3 := \{(\nu,x) \in M^1(\mathbb{H}) \times \mathbb{H}^\infty : \nu_{|1|} = \infty\}.$$

Then by Theorem 2.24, Lemma 2.20, and the (Kolmogorov) strong law of large numbers for $(\mathbb{R}^3, +)$, A is a limit statute. Now Corollary 2.4 follows from Theorem 2.24, Proposition 2.4, and Lemma 2.19. □

In 3.2.1 we will apply the subsequence principle also to the Marcinkiewicz Zygmund strong laws of large numbers.

Chapter 3

Other limit theorems on $\mathbb{H}$

3.1 Weak theorems

3.1.1 Universal laws

In this section we will prove the following complement to Theorem 1.1 (cf. Neuenschwander (1995e)):

Theorem 3.1 *Let $A \in \mathrm{Aut}(\mathbb{H})$. Then a probability measure $\nu \in M^1(\mathbb{H})$ is A-universal iff $^\circ\nu$ is $^\circ A$-universal on $(\mathbb{R}^3, +)$.*

This is an analogue of Remark 3.4 (which follows from our Propositions 1.2, 1.3, and the identification $\mathcal{A} \leftrightarrow^\circ \mathcal{A}$) in Hazod (1993) for the case where we allow also shifts x_k.
Proof of Theorem 3.1: 1. We first prove necessity. Assume $\nu \in M^1(\mathbb{H})$ is A-universal, i.e., for a given c.c.s. $\{\mu_t\}_{t>0}$ on $\mathbb{H}$, there exist subsequences $\{n_k\}_{k\geq 1}$, $\{m_k\}_{k\geq 1} \subset \{n\}_{n>1}$ such that

$$A^{m_k}(\nu * \varepsilon_{x_k})^{*[n_k t]} \xrightarrow{w} \mu_t \quad (k \to \infty) \quad (t \geq 0).$$

It is easy to see that from Proposition 1.3 and Siebert (1981), Propositions 6.1 and 6.4, it follows that

$$p(A^{m_k}(x_k)) \to 0 \quad (k \to \infty),$$
$$p(A^{m_k}(\nu)) \xrightarrow{w} \varepsilon_0 \quad (k \to \infty),$$

and thus

$$q(A^{m_k}(x_k)) \to 0 \quad (k \to \infty),$$

hence

$$A^{m_k}(x_k) \to 0 \quad (k \to \infty). \tag{3.1}$$

From now on, the proof of the necessity part is analogous to the proof of Lemma 2.4.
2. Now we come to the sufficiency part. Suppose $^\circ\nu \in M^1(\mathbb{R}^3)$ is A-universal and let $\{\mu_t\}_{t>0} = \{\mathrm{Exp}\, t\mathcal{A}\}_{t>0}$ be an arbitrary c.c.s. on $\mathbb{H}$. By Proposition 1.3 there are subsequences $\{m_k\}_{k>1}, \{n_k\}_{h\geq 1} \subset \{n\}_{n>1}$ such that

$$\mathrm{Exp}\, t^\circ \mathcal{A}_k \xrightarrow{w} \mathrm{Exp}\, t^\circ \mathcal{A} \quad (k \to \infty) \quad (t \geq 0), \tag{3.2}$$

where

$$^\circ\!\mathcal{A}_k(^\circ f) = n_k \int_{\mathbb{H}} (^\circ f(^\circ x) - {}^\circ f(0))^\circ A^{m_k}(^\circ \nu * \varepsilon_{^\circ x_k})(d^\circ x).$$

By (3.2) and the analogue of (2.17)

$$\operatorname{Exp} t^\circ \tilde{\mathcal{A}}_k \xrightarrow{w} \operatorname{Exp} t^\circ \mathcal{A} \quad (k \to \infty) \quad (t \geq 0),$$

where

$$\tilde{\mathcal{A}} = n_k \int_{\mathbb{H}} (f(x) - f(0)) A^{m_k}(\nu * \varepsilon_{x_k})(dx),$$

so by Proposition 1.2 and the identification $\mathcal{A} \leftrightarrow^\circ \mathcal{A}$ we have

$$\operatorname{Exp} t\tilde{\mathcal{A}}_k \xrightarrow{w} \operatorname{Exp} t\mathcal{A} \quad (k \to \infty) \quad (t \geq 0),$$

which, by Proposition 1.3, yields sufficiency.□

Remark 3.1 *Theorem 3.1 and the analogue of its proof remain true if "A-universal" is replaced by "universal".*

3.1.2 Domains of attraction of stable semigroups

First of all, we mention that the proof of Theorem 3.1 and the corresponding result for euclidean spaces yields the following:

Theorem 3.2 *The c.c.s. $\{\mu_t\}_{t\geq 0}$ on $\mathbb{H}$ is $\{t^A\}_{t>0}$-stable ($\{t^A\}_{t>0} \subset Aut(\mathbb{H})$) iff $DOA(\{t^A\}_{t>0}, \{\mu_t\}_{t\geq 0}) \neq \emptyset$ iff $DONA(\{t^A\}_{t>0}, \{\mu_t\}_{t>0}) \neq \emptyset$.*

Now we continue to present Scheffler's (1993) work on domains of attraction, which we began in 2.1.3. In contrast to 2.1.3, here the case of c.c.s. without Gaussian component is being considered.
Let $\mathcal{B}$ be as in 2.1.3. Let $\{t^A\}_{t>0} = \{\sigma_{M,m}(t)\}_{t>0}$ (cf. 2.4). The following is a slight variant of Scheffler (1993), Theorem 5.1:

Theorem 3.3 *Suppose $\{\mu_t\}_{t\geq 0}$ is an L-S-full strictly $\{\sigma_{M,m}(t)\}_{t>0}$-stable semigroup on $\mathbb{H}$ without Gaussian component and with Lévy measure η. Let $\nu \in M^1(\mathbb{H})$. Consider the conditions*
(i) $\nu \in SDOA(\{\mu_t\}_{t\geq 0}, \mathcal{B})$,
(ii) there exists $\{\tau_n\}_{n\geq 1} \subset \mathcal{B}$ such that for every Borel neighborhood U of 0

$$n(\tau_n(\nu))|_{\mathrm{cpl}\,U} \xrightarrow{w} \eta|_{\mathrm{cpl}\,U} \quad (n \to \infty). \tag{3.3}$$

Then we have (i) $\Longrightarrow$ (ii) and, if ν and μ_t are symmetric, then also (ii) $\Longrightarrow$ (i).

Proof: (i)$\Longrightarrow$ (ii): This follows just from Proposition 1.3 and Siebert (1981), Proposition 6.4.
(ii) $\Longrightarrow$ (i): It follows from (3.3) with the aid of Meerschaert (1986), Theorem that

$$p(\nu) \in SDOA(\{p(\mu_t)\}_{t\geq 0}, \{mF : F \text{ [skew-]symplectic}, m > 0\}).$$

Furthermore, by Scheffler (1994), Theorem 3.1 we have

$$\limsup_{n\to\infty} n \int_{||p(x)||<\varepsilon} ||p(x)||^2 p(\tau_n(\nu))(dx) \to 0 \quad (\varepsilon \to 0).$$

So by Scheffler (1994), Theorem 3.1 it remains to verify

$$\limsup_{n\to\infty} n \int_{q(x)<\varepsilon} q(x)^2 q(\tau_n(\nu))(dx) \to 0 \quad (\varepsilon \to 0). \tag{3.4}$$

Write $\theta_a(x) := ax$ $(x \in I\!R, a \geq 0)$. We have

$$n\theta_{m_n^2}(q(\nu))|_{\mathrm{cpl}\,U} \xrightarrow{w} q(\eta)|_{\mathrm{cpl}\,U} \quad (n \to \infty) \tag{3.5}$$

for every Borel neighborhood U of 0 (where

$$\tau_n = \begin{pmatrix} m_n F_n & 0 \\ 0 & |-|m_n^2 \end{pmatrix},$$

F_n [skew-]symplectic, $m_n > 0$). Let $\mathcal{A} = (\xi, 0, \eta)$ be the generating distribution of $\{\mu_t\}_{t>0}$. Then $q(\mathcal{A}) = (q(\xi), 0, q(\eta))$ generates a strictly $\{t^{2m}\}_{t>0}$-stable c.c.s. of symmetric measures on $I\!R$. The case where $q(\eta) \equiv 0$ is trivial: here by (3.5)

$$n\theta_{m_n^2}(q(\nu))|_{\mathrm{cpl}\,U} \xrightarrow{w} 0|_{\mathrm{cpl}\,U} \quad (n \to \infty)$$

for every Borel neighborhood U of 0, which readily implies (3.4). In the case $q(\eta) \not\equiv 0$, $\{\mathrm{Exp}\, tq(\mathcal{A})\}_{t\geq 0}$ is an L-S-full c.c.s. without Gaussian component and the assertion follows from Meerschaert (1986), Theorem and Scheffler (1994), Theorem 3.1. □
By looking at a suitable desintegration of the Lévy measure, Theorem 3.3 implies a criterion for the strict domain of normal attraction (cf. Scheffler (1993), Corollary 5.5). For general simply connected nilpotent Lie groups, the $\{t^A\}_{t>0}$-domain of attraction of c.c.s. without Gaussian component has been characterized in Neuenschwander (1995c) (see also Carnal (1986)). It was possible to describe it by an analogue of the classical Doeblin-Gnedenko conditions (involving slow variation) (see e.g. Breiman (1968), Theorem 9.34).

3.1.3 Lightly trimmed products

By analogy with 2.1.4, this section deals with so-called lightly trimmed products. It presents the main theorem of the author's Ph.D. thesis (cf. Neuenschwander (1991, 1995d)).
We will have to use the following lemma, which is similar to Billingsley (1968), Theorem 4.2:

Lemma 3.1 *Assume $Z_1, Z_2, \ldots$ are $I\!R^d$-valued random variables with a decomposition $Z_n = A_n(\varepsilon) + D_n(\varepsilon)$ $(n \geq 1)$ for every $\varepsilon > 0$ such that*

$$\limsup_{n\to\infty} P(||D_n(\varepsilon)|| \geq \delta) \to 0 \quad (\varepsilon \to 0) \tag{3.6}$$

for all $\delta > 0$ and

$$\mathcal{L}(A_n(\varepsilon)) \xrightarrow{w} \mathcal{L}(A(\varepsilon)) \quad (n \to \infty)$$

for any $\varepsilon > 0$ and certain ($\mathbb{R}^d$-valued) random variables $A(\varepsilon)$. Then there is an ($\mathbb{R}^d$-valued) random variable Z such that

$$\mathcal{L}(Z_n) \xrightarrow{w} \mathcal{L}(Z). \tag{3.7}$$

Proof: Denote by $\rho(\nu, \mu)$ the Prohorov distance between $\nu, \mu \in M^1(\mathbb{R}^d)$ (cf. Araujo, Giné (1980), p.12). It is well-known that ρ metrizes the weak topology (cf. Araujo, Giné (1980), Theorem 1.2.7) and that the resulting metric space is complete (cf. Araujo, Giné (1980), Exercise 17 on p.18). Furthermore, let the distance in probability between the $\mathbb{R}^d$-valued random variables X, Y be defined as

$$d_{pr}(X, Y) := \inf\{\delta \geq 0 : P(||X - Y|| > \delta) \leq \delta\}.$$

We get by Araujo, Giné (1980), Exercise 14(iii) on p.18 (see also Hahn, Hudson, Veeh (1989), p.20)

$$\begin{aligned}
\rho(\mathcal{L}(Z_n), \mathcal{L}(Z_m)) &\leq \rho(\mathcal{L}(Z_n), \mathcal{L}(A_n(\varepsilon))) + \rho(\mathcal{L}(A_n(\varepsilon)), \mathcal{L}(A_m(\varepsilon))) \\
&\quad + \rho(\mathcal{L}(A_m(\varepsilon)), \mathcal{L}(Z_m)) \\
&\leq d_{pr}(D_n(\varepsilon), 0) + \rho(\mathcal{L}(A_n(\varepsilon)), \mathcal{L}(A_m(\varepsilon))) \\
&\quad + d_{pr}(D_m(\varepsilon), 0),
\end{aligned}$$

which entails, together with assumption (3.6), that $\{\mathcal{L}(Z_n)\}_{n \geq 1}$ is a Cauchy sequence with respect to ρ. Hence (3.7) follows. □

Remark 3.2 *Clearly,*

$$\mathcal{L}(A(\varepsilon)) \xrightarrow{w} \mathcal{L}(Z) \quad (\varepsilon \to 0).$$

By $X_{n:1} \leq X_{n:2} \leq \dots \leq X_{n:n}$ we denote the ascending order statistics of the real-valued random variables $X_1, X_2, \dots, X_n$.
The following property of exponential random variables is well-known:

Lemma 3.2 *Let $E_1, E_2, \dots, E_{n+1}$ be i.i.d. random variables obeying to an exponential law with parameter $\lambda > 0$ and let $U_1, U_2, \dots, U_n$ be i.i.d. with uniform law on the interval $[0, 1]$. Assume $\{U_j\}_{1 \leq j \leq n}$ is independent of $\{E_j\}_{1 \leq j \leq n+1}$, and put*

$$\Gamma_i := \sum_{j=1}^{i} E_j \quad (1 \leq i \leq n + 1).$$

Then we have

$$\mathcal{L}(\Gamma_1, \Gamma_2, \dots, \Gamma_n) = \mathcal{L}(U_{n:1}\Gamma_{n+1}, U_{n:2}\Gamma_{n+1}, \dots, U_{n:n}\Gamma_{n+1}).$$

Let μ be some full $\{t^A\}_{t>0}$-stable measure on $(\mathbb{R}^d, +)$ without Gaussian component and let h denote the Haar measure on the invariance group $S(\mu)$ of μ. Then

$$\operatorname{Spec} A \subset \{z \in \mathbb{C} : \operatorname{Re} z > \frac{1}{2}\} \tag{3.8}$$

(cf. Hahn, Hudson, Veeh (1989), p.4, line 31ff.) and

$$(x, y) := \int_{S(\mu)} \int_0^1 \langle M t^A x, M t^A y \rangle \frac{dt}{t} h(dM)$$

defines an inner product on $\mathbb{R}^d$ (cf. Hahn, Hudson, Veeh (1989), (2.1), Hudson, Jurek, Veeh (1986)). Let S denote the unit sphere with respect to this inner product. Then the map

$$\begin{array}{rcl}]0, \infty[\times S & \to & \mathbb{R}^d \backslash \{0\} \\ (t, y) & \mapsto & t^A y \end{array} \tag{3.9}$$

is a homeomorphism (cf. Hahn, Hudson, Veeh (1989), p.5). The Lévy measure of μ then has the form

$$\eta(dx) = \lambda \nu(dy) \frac{dt}{t^2} \quad (\lambda \geq 0, \nu \in M^1(S)) \tag{3.10}$$

(cf. Hahn, Hudson, Veeh (1989), Theorem 1).

The following theorem tells us the following: Let X be a symmetric random variable with values in $\mathbb{H}$ and $\mathcal{S} = \{\operatorname{Exp} t\mathcal{A}\}_{t>0}$ some $\{t^A\}_{t>0}$-stable semigroup on $\mathbb{H}$ with symmetric Lévy measure, without Gaussian component, and such that ${}^\circ\mathcal{A}$ generates a full c.c.s. on $(\mathbb{R}^3, +)$. Let $\{X_n\}_{n \geq 1}$ be i.i.d. copies of X and assume $\mathcal{L}(X) \in DOA(\mathcal{S}, \{t^A\}_{t>0})$ with norming sequence $\{(t_n^A, x_n)\}_{n \geq 1} \subset Aut(\mathbb{H}) \times \mathbb{H}$. Then the lightly trimmed products

$$P_n^{(k)} := t_n^A\Big(\prod_{j \in J_n^k} (X_j \cdot x_n)\Big),$$

which arise from the non-trimmed products

$$P_n^{(k)} := t_n^A\Big(\prod_{j=1}^{n} (X_j \cdot x_n)\Big)$$

by skipping those indexes j for which the "radial component" T_j with respect to the polar coordinate decomposition (3.9) (i.e. $X_j = T_j^A Y_j$ $(T_j \geq 0, Y_j \in S)$) is among the k largest values of $T_1, T_2, \ldots, T_n$ (roughly speaking), are convergent, too. The limit may be described in terms of order statistics of i.i.d. uniform (real-valued) random variables and independent random variables on S distributed according to the law ν in the desintegration (3.10) of the Lévy measure of $\mathcal{S}$. This is in some sense a generalization of the so-called "Lévy construction of second kind" for stable measures. Such an assertion for the real line was proved in Janssen (1989), Theorem 5.2, where it is shown that the limit can be obtained from the "Lévy construction of first kind" (cf. Carnal (1986), also for the group case) by ignoring the k first jumps of the Poisson process involved.

Let A be a derivation of the Lie algebra of $\mathbb{H}$ and $\mathcal{S} = \{\operatorname{Exp} t\mathcal{A}\}_{t>0}$ a $\{t^A\}_{t>0}$-stable semigroup on $\mathbb{H}$. Assume ${}^\circ\mathcal{A}$ generates a full c.c.s. on $(\mathbb{R}^3, +)$ and $\mathcal{A} = (\gamma, 0, \eta)$, where η is as in (3.10). Let $\{X_j\}_{j \geq 1}$ be i.i.d. $\mathbb{H}$-valued random variables defined on some common probability space $(\Omega, \mathcal{F}, P)$ such that $\mathcal{L}(X_1) \in DOA(\mathcal{S}, \{t^A\}_{t>0})$

with norming sequence $\{(t_n^A, x_n)\}_{n\geq 1} \subset Aut(\mathbb{H}) \times \mathbb{H}$. Assume the law of X_1 and ν are symmetric. Consider the polar coordinate decomposition (3.9) for X_j, i.e. $X_j = T_j^A Y_j$ $(T_j \geq 0, Y_j \in S)$. It follows from (3.8), Lemma 2.4, the form of the centering constants in the general central limit theorem (Araujo, Giné (1980), Theorem 3.5.9), which are zero by the symmetry of $\mathcal{L}(X_1)$, and the convergence of types theorem on $(\mathbb{R}^3, +)$ that the x_n have to be of the form

$$x_n = \frac{1}{n} t_n{}^A(\gamma + \Delta_n)$$

with $\Delta_n \in \mathbb{R}^3$ and

$$\Delta_n \to 0 \quad (n \to \infty).$$

Let $\Gamma_1 < \Gamma_2 < \dots$ denote the jump-times of a Poisson process $\{N(t)\}_{t\geq 0}$ with parameter λ, $\{Z_j\}_{j\geq 1}$ a process (independent of $\{N(t)\}_{t\geq 0}$) of i.i.d. random variables on S distributed according to the law ν and $\{U_j^{(n)}\}_{n\geq 1; 1\leq j\leq n}$ an array of random variables which is independent of $\{Z_j\}_{j\geq 1}$ and such that $U_1^{(n)}, U_2^{(n)}, \dots, U_n^{(n)}$ are i.i.d. uniformly distributed over the interval $[0, \frac{n}{\lambda}]$. Define a process which is independent of $\{\{N(t)\}_{t\geq 0}, \{Z_j\}_{j\geq 1}\}$ and consists of i.i.d. random variables $\{V_j\}_{j\geq 1}$ which are uniformly distributed on $[0,1]$. Let the permutation σ_n of $\{1, 2, \dots, n\}$ be defined as follows (cf. Hahn, Hudson, Vech (1989), p.6): Put $\sigma_n(j) = i$ if T_j is the i-th largest element among $T_1, T_2, \dots, T_n$ (in case of ties introduce an additional independent random permutation of the indexes constituting the tie such that all permutations σ_n of $\{1, 2, \dots, n\}$ are equally likely). Consider the ordered index sets

$$J_n^k := \{j \in \{1, 2, \dots, n\} : \sigma_n(j) > k\}$$
(ordered by the natural ordering of $\mathbb{N}$)

and

$$I_{\infty,\varepsilon,\rho}^k := \{j \geq k+1 : \Gamma_j^{-1} \in [\varepsilon, \rho]\}$$
(ordered by a random permutation which is independent of $\{\{N(t)\}_{t\geq 0}, \{V_j\}_{j\geq 1}, \{Z_j\}_{j\geq 1}\}$).

Similarly, let I_n^k be the ordered index set which arises by replacing $T_1, T_2, \dots, T_n$ by $U_1^{(n)}, U_2^{(n)}, \dots, U_n^{(n)}$ in the above definition of J_n^k (here ties occur with probability zero). The ordering plays a role when multiplying random variables indexed by one of the above sets. Let

$$N(\varepsilon, \rho, k) := \# I_{\infty,\varepsilon,\rho}^k$$

and consider the products /sums

$$P_n^{(k)} := t_n^A\Big(\prod_{j\in J_n^k} (X_j \cdot x_n)\Big),$$
$$P_{n,B}^{(k)} := t_n^A\Big(\prod_{j\in J_n^k} (X_j 1_B(t_n T_j) \cdot x_n)\Big),$$

$$S^{(k)}_{n,B} := t_n^{A}\Big(\sum_{j\in J_n^k}(X_j \mathbf{1}_B(t_nT_j) + x_n)\Big),$$

$$\overline{P}_{n,B} := t_n^{A}\Big(\prod_{j=1}^{n} X_j \mathbf{1}_B(t_nT_j)\Big),$$

$$\overline{S}^{(k)}_{n,B} := t_n^{A}\Big(\prod_{j\in J_n^k} X_j \mathbf{1}_B(t_nT_j)\Big),$$

$$Q^{(k)}_{n} := \prod_{j\in I_n^k}\Big((U_j^{(n)})^{-A} Z_j \cdot \frac{\gamma}{n}\Big),$$

$$Q_{n,t} := \prod_{j=1}^{\lfloor nt\rfloor}\Big((U_j^{(n)})^{-A} Z_j \cdot \frac{\gamma}{n}\Big),$$

$$Q^{(k)}_{n,B} := \prod_{j\in I_n^k}\Big((U_j^{(n)})^{-A} Z_j \mathbf{1}_B((U_j^{(n)})^{-1}) \cdot \frac{\gamma}{n}\Big),$$

$$R_{n,t} := \sum_{j=1}^{\lfloor nt\rfloor}\Big((U_j^{(n)})^{-A} Z_j + \frac{\gamma}{n}\Big),$$

$$\tilde{R}_{n,t} := \sum_{j=1}^{\lfloor nt\rfloor}\Big((U_j^{(n)})^{-A} Z_j \cdot \frac{\gamma}{n}\Big),$$

$$\overline{R}_{n,t} := \sum_{j=1}^{\lfloor nt\rfloor}(U_j^{(n)})^{-A} Z_j$$

(where $B \subset [0,\infty[$), as well as (for $0 < \varepsilon < \rho$) the difference

$$D^{(k)}_{n,\varepsilon,\rho} := P^{(k)}_{n,]0,\rho[} - P^{(k)}_{n,]\varepsilon,\rho[}.$$

Theorem 3.4 *Under the afore-mentioned conditions, $\mathcal{L}(P_n^{(k)})$ is weakly convergent to some random variable $P^{(k)}_{\infty}$, which is also the weak limit of $\{\mathcal{L}(Q_n^{(k)})\}_{n\geq 1}$.*

Proof: Roughly speaking, the plan of the proof is as follows: First, we prove weak convergence of the truncated products $P^{(k)}_{n,]\varepsilon,\rho[}$ and then we estimate the differences $P^{(k)}_{n,]0,\rho[} - P^{(k)}_{n,]\varepsilon,\rho[}$ and $P^{(k)}_{n} - P^{(k)}_{n,]0,\rho[}$ in order to apply Lemma 3.1. Relying on Lemma 2.4, this can be done by making use of well-known properties of triangular arrays of probability measures on $(\mathbb{R}^3, +)$. A similar argument holds for $Q_n^{(k)}$ instead of $P_n^{(k)}$. The proof is related to the corresponding arguments in Le Page, Woodroofe, Zinn (1981) and Hahn, Hudson, Veeh (1989). However, we will not only truncate from below, but from above, too. Let $0 < \varepsilon < \rho$.

1. With the aid of (3.8), Lemma 2.4, and Hahn, Hudson, Veeh (1989), Remark 1 one finds

$$\mathcal{L}(P^{(k)}_{n,]\varepsilon,\rho[}) \xrightarrow{w} \mathcal{L}(P^{(k)}_{\infty,]\varepsilon,\rho[}) \quad (n \to \infty). \tag{3.11}$$

where

$$P^{(k)}_{\infty,]\varepsilon,\rho[} := \prod_{j\in I^k_{\infty,\varepsilon,\rho}} \tilde{V}_{N(\varepsilon,\rho,k),j}\gamma \cdot \Gamma^{-A}_{\sigma(j)} Z_{\sigma(j)} \cdot (1 - V_{N(\varepsilon,\rho,k):N(\varepsilon,\rho,k)})\gamma,$$

$$\{\tilde{V}_{N(\varepsilon,\rho,k),j}\}_{j\in I^k_{\infty,\varepsilon,\rho}} := (V_{N(\varepsilon,\rho,k):1}, V_{N(\varepsilon,\rho,k):2} - V_{N(\varepsilon,\rho,k):1}, \ldots, V_{N(\varepsilon,\rho,k):N(\varepsilon,\rho,k)} - V_{N(\varepsilon,\rho,k):N(\varepsilon,\rho,k)-1})$$

(The idea is the same as that of the proof of relation (10) in Le Page, Woodroofe, Zinn (1981) using Hahn, Hudson, Veeh (1989), Remark 1 instead of Lemmas 1, 2 in Le Page, Woodroofe, Zinn (1981). Furthermore, recall that by the definition of $P^{(k)}_{n,B}$ every $j \in J^k_n$ yields an x_n (and thus a contribution to a fractional portion of γ); by the random ordering of $T_1, T_2, \ldots, T_n$, the above representation follows.)

2. For every $\delta > 0$ we have by (3.8), Lemma 2.4, and Hahn, Hudson, Veeh (1989), Remark 1

$$\begin{aligned}\limsup_{n\to\infty} P(\|P^{(k)}_n - P^{(k)}_{n,[0,\rho[}\| \geq \delta) &\leq \limsup_{n\to\infty} P(t_n T_{\sigma_n^{-1}(k-1)} \geq \rho)\\ &= P(\Gamma_{k-1}^{-1} \geq \rho)\\ &\to 0 \quad (\rho \to \infty).\end{aligned} \tag{3.12}$$

3. Now we estimate $D^{(k)}_{n,\varepsilon,\rho}$. We calculate

$$D^{(k)}_{n,\varepsilon,\rho} = \overline{S}^{(k)}_{n,]0,\varepsilon[} + \frac{1}{2}A^{(k)}_{n,\varepsilon,\rho} + \frac{1}{2}B^{(k)}_{n,\varepsilon} + \frac{1}{2}C^{(k)}_{n,\varepsilon,\rho}, \tag{3.13}$$

where

$$A^{(k)}_{n,\varepsilon,\rho} := [S^{(k)}_{n,]0,\rho[}, \frac{1}{n}(\gamma + \Delta_n)] - [S^{(k)}_{n,]\varepsilon,\rho[}, \frac{1}{n}(\gamma + \Delta_n)],$$

$$B^{(k)}_{n,\varepsilon} := t_n^A\Big(\sum_{\substack{1<i<j<n\\ i,j\in J^k_n}} [X_i\mathbf{1}_{]0,\varepsilon[}(t_nT_i), X_j\mathbf{1}_{]0,\varepsilon[}(t_nT_j)]\Big),$$

$$\begin{aligned}C^{(k)}_{n,\varepsilon,\rho} := t_n^A\Big(\sum_{\substack{1<i<j<n\\ i,j\in J^k_n}} &([X_i\mathbf{1}_{]0,\varepsilon[}(t_nT_i),\\ &X_j\mathbf{1}_{[\varepsilon,\rho[}(t_nT_j) + \frac{1}{n}t_n{}^A(\gamma + \Delta_n)]\\ &+[X_i\mathbf{1}_{[\varepsilon,\rho[}(t_nT_i) + \frac{1}{n}t_n{}^A(\gamma + \Delta_n), X_j\mathbf{1}_{]0,\varepsilon[}(t_nT_j)])\Big).\end{aligned} \tag{3.14}$$

From (3.8), Lemma 2.4, the symmetry of $\mathcal{L}(X_1)$, and Araujo, Giné (1980), Theorem 3.5.6 (1) $\Rightarrow$(3) (its premise concerning the Lévy measure follows from (2.12) (which is valid also here by the same reason ((3.8))), Propositions 1.2 and 1.3, and the identification $\mathcal{A} \leftrightarrow^{\circ} \mathcal{A}$ by substituting functions in the generating distributions which vanish in a neighborhood of 0) it follows that $\{\mathcal{L}(\overline{S}^{(0)}_{n,]0,\rho[})\}_{n\geq 1}$ and $\{\mathcal{L}(\overline{S}^{(0)}_{n,]\varepsilon,\rho[})\}_{n>1}$ are weakly relatively compact; by the finite truncation from above (with ρ), the same holds for $\{\mathcal{L}(\overline{S}^{(k)}_{n,]0,\rho[})\}_{n\geq 1}$ and $\{\mathcal{L}(\overline{S}^{(k)}_{n,]\varepsilon,\rho[})\}_{n>1}$; now we conclude

$$\mathcal{L}(A^{(k)}_{n,\varepsilon,\rho}) \xrightarrow{w} \varepsilon_0 \quad (n \to \infty). \tag{3.15}$$

By (3.8), Lemma 2.4, and Hahn, Hudson, Veeh (1989), Remark 1 we get

$$\limsup_{n\to\infty} P(t_nT_{\sigma_n^{-1}(k)} < \varepsilon) = P(\Gamma_k^{-1} < \varepsilon) \to 0 \quad (\varepsilon \to 0). \tag{3.16}$$

As for $B^{(k)}_{n,\varepsilon}$, observe first that

$$B^{(0)}_{n,\varepsilon} = 2(\overline{P}_{n,[0,\varepsilon]} - \overline{S}^{(0)}_{n,[0,\varepsilon]}). \tag{3.17}$$

By (2.12), (3.16), (3.17), and Propositions 1.2, 1.3, identifying $\mathcal{A}$ and ${}^{\circ}\mathcal{A}$, it follows that

$$\begin{aligned}
&\limsup_{n\to\infty} P(\|B^{(k)}_{n,\varepsilon}\| \geq \delta) \\
&\leq \limsup_{n\to\infty} P(t_n T_{\sigma_n^{-1}(k)} < \varepsilon) + \limsup_{n\to\infty} P(\|B^{(0)}_{n,\varepsilon}\| \geq \delta) \\
&\leq \limsup_{n\to\infty} P(t_n T_{\sigma_n^{-1}(k)} < \varepsilon) + \limsup_{n\to\infty} P(\|\overline{P}_{n,[0,\varepsilon]}\| \geq \frac{\delta}{4}) \\
&\quad + \limsup_{n\to\infty} P(\|\overline{S}^{(0)}_{n,[0,\varepsilon]}\| \geq \frac{\delta}{4}) \\
&= P(\Gamma_k^{-1} < \varepsilon) + (\mathrm{Exp}\mathcal{A}_\varepsilon)(\{x \in I\!H : \|{}^{\circ}x\| \geq \frac{\delta}{4}\}) \\
&\quad + (\mathrm{Exp}{}^{\circ}\mathcal{A}_\varepsilon)(\{x \in I\!R^3 : \|x\| \geq \frac{\delta}{4}\}),
\end{aligned} \tag{3.18}$$

where

$$\begin{aligned}
\mathcal{A}_\varepsilon &:= (0, 0, \eta_\varepsilon), \\
\eta_\varepsilon(dx) &:= \begin{cases} \lambda\nu(dy)\frac{dt}{t^2} & : \ 0 < t < \varepsilon \\ 0 & : \ t \geq \varepsilon \end{cases} \\
&\quad (x = t^A y \in I\!H\setminus\{0\}, t > 0, y \in S).
\end{aligned} \tag{3.19}$$

Since $\mathcal{A}_\varepsilon(f) \to 0$ $(\varepsilon \to 0)$ $(f \in C_b^\infty(I\!H))$, we thus obtain by (3.18), Propositions 1.2, 1.3, and the identification $\mathcal{A} \leftrightarrow {}^{\circ}\mathcal{A}$

$$\limsup_{n\to\infty} P(\|B^{(k)}_{n,\varepsilon}\| \geq \delta) \to 0 \quad (\varepsilon \to 0). \tag{3.20}$$

Similarly

$$\begin{aligned}
&\limsup_{n\to\infty} P(\|\overline{S}^{(k)}_{n,[0,\varepsilon]}\| \geq \delta) \\
&\leq \limsup_{n\to\infty} P(t_n T_{\sigma_n^{-1}(k)} < \varepsilon) + \limsup_{n\to\infty} P(\|\overline{S}^{(0)}_{n,\varepsilon}\| \geq \delta) \\
&= P(\Gamma_k^{-1} < \varepsilon) + (\mathrm{Exp}{}^{\circ}\mathcal{A}_\varepsilon)(\{x \in I\!R^3 : \|x\| \geq \delta\})
\end{aligned}$$

and thus

$$\limsup_{n\to\infty} P(\|\overline{S}^{(k)}_n\| \geq \delta) \to 0 \quad (\varepsilon \to 0). \tag{3.21}$$

Now we turn to the estimation of $C^{(k)}_{n,\varepsilon,\rho}$. Put

$$\delta_{ij} := \begin{cases} 1 & : \ i < j \\ 0 & : \ i = j \\ -1 & : \ i > j. \end{cases}$$

With this we may write

$$C_{n,\varepsilon,\rho}^{(k)} := t_n^A(\sum_{i\in J_n^k}[X_i\mathbf{1}_{]0,\varepsilon]}(t_nT_i), \sum_{j\in J_n^k}\delta_{ij}(X_j\mathbf{1}_{]\varepsilon,\rho]}(t_nT_j) + \frac{1}{n}t_n{}^A(\gamma+\Delta_n))]). \tag{3.22}$$

Hence

$$C_{n,\varepsilon,\rho}^{(0)} = C_{n,\varepsilon,\rho}^{(k)} + F_{1,n,\varepsilon,\rho}^{(k)} - F_{2,n,\varepsilon,\rho}^{(k)} + F_{3,n,\varepsilon,\rho}^{(k)}. \tag{3.23}$$

$$F_{1,n,\varepsilon,\rho}^{(k)} = t_n^A(\sum_{i=1}^{n}[X_i\mathbf{1}_{]0,\varepsilon]}(t_nT_i), \sum_{j\notin J_n^k}\delta_{ij}(X_j\mathbf{1}_{]\varepsilon,\rho]}(t_nT_j) + \frac{1}{n}t_n{}^A(\gamma+\Delta_n))]).$$

$$F_{2,n,\varepsilon,\rho}^{(k)} = t_n^A(\sum_{i\notin J_n^k}[X_i\mathbf{1}_{]0,\varepsilon]}(t_nT_i), \sum_{j\notin J_n^k}\delta_{ij}(X_j\mathbf{1}_{]\varepsilon,\rho]}(t_nT_j) + \frac{1}{n}t_n{}^A(\gamma+\Delta_n))]).$$

$$F_{3,n,\varepsilon,\rho}^{(k)} = t_n^A(\sum_{i\notin J_n^k}[X_i\mathbf{1}_{]0,\varepsilon]}(t_nT_i), \sum_{j=1}^{n}\delta_{ij}(X_j\mathbf{1}_{]\varepsilon,\rho]}(t_nT_j) + \frac{1}{n}t_n{}^A(\gamma+\Delta_n))]).$$

For $h = 2,3$ we get by (3.16)

$$\limsup_{n\to\infty} P(||F_{h,n,\varepsilon,\rho}||^{(k)} \geq \delta) \leq \limsup_{n\to\infty} P(t_nT_{\sigma_n^{-1}(k)} < \varepsilon) \to 0 \quad (\varepsilon\to 0). \tag{3.24}$$

As far as $F_{1,n,\varepsilon,\rho}^{(k)}$ is concerned, one easily obtains (using the symmetry of $\mathcal{L}(X_1)$ and the additivity of the variance of a sum of independent random variables)

$$E(X_i\mathbf{1}_{]0,\varepsilon]}(t_nT_i)|\{X_j\mathbf{1}_{]\varepsilon,\rho]}(t_nT_j)\}_{1\leq j\leq n}) = 0 \quad (1\leq i\leq n). \tag{3.25}$$

$$\begin{aligned} &E(||F_{1,n,\varepsilon,\rho}^{(k)}||^2|\{X_j\mathbf{1}_{]\varepsilon,\rho]}(t_nT_j)\}_{1\leq j\leq n}) \\ &\leq C(\rho)nE(||t_n^A(X_n\mathbf{1}_{]0,\varepsilon]}(t_nT_n))||^2)/P(t_nT_n<\varepsilon). \end{aligned} \tag{3.26}$$

By (2.12), (3.10), and Propositions 1.2, 1.3 we get

$$nP(t_nT_n\geq\varepsilon) \to \frac{1}{\varepsilon} \quad (n\to\infty),$$

hence

$$P(t_nT_n<\varepsilon)\to 1 \quad (n\to\infty). \tag{3.27}$$

So by (3.26)-(3.28), Markov's inequality, Lemma 2.4. and Araujo, Giné (1980), Theorem 3.5.9 (ii) (where by the symmetry of $\mathcal{L}(X_1)$ the centering constant in (ii) drops and by the assumption of no Gaussian component the limit is 0) we may conclude that

$$\limsup_{n\to\infty} P(||F_{1,n,\varepsilon,\rho}^{(k)}||\geq\delta)\to 0 \quad (\varepsilon\to 0). \tag{3.28}$$

So it remains to consider $C_{n,\varepsilon,\rho}^{(0)}$, which can be treated similarly as $F_{1,n,\varepsilon,\rho}^{(k)}$. One obtains (as above)

$$\begin{aligned} &E(||C_{n,\varepsilon,\rho}^{(0)}||^2|\{X_j\mathbf{1}_{]\varepsilon,\rho]}\}_{1\leq j\leq n}) \\ &\leq C\frac{1}{n}\sum_{i=1}^{n}||t_n^A(\sum_{j=1}^{n}\delta_{ij}(X_j\mathbf{1}_{]\varepsilon,\rho]}(t_nT_j) + \frac{1}{n}t_n{}^A(\gamma+\Delta_n)))||^2 \\ &\cdot nE||t_n^A(X_n\mathbf{1}_{]0,\varepsilon]}(t_nT_n))||^2/P(t_nT_n<\varepsilon); \end{aligned} \tag{3.29}$$

however by (3.8), Lemma 2.4, Rvačeva (1962), Lemma 2.4, and the symmetry of $\mathcal{L}(X_1)$

$$\begin{aligned}\limsup_{n\to\infty} E\|t_n^A(\sum_{j=1}^n \delta_{ij}(X_j \mathbf{1}_{[\varepsilon,\rho]}(t_n T_j)) + \frac{1}{n} t_n^{-A}(\gamma+\Delta_n)))\|^2 \\ = \limsup_{n\to\infty} nE\|t_n^A(X_n \mathbf{1}_{[\varepsilon,\rho]}(t_n T_n))\|^2 + \|\gamma\|^2 \\ = D(\rho) + \|\gamma\|^2. \end{aligned} \tag{3.30}$$

and thus again by (3.27), (3.29), (3.30), Markov's inequality, (3.8), Lemma 2.4, the symmetry of $\mathcal{L}(X_1)$, and Araujo, Giné (1980), Theorem 3.5.9 (as above)

$$\limsup_{n\to\infty} P(\|C_{n,\varepsilon,\rho}^{(0)}\| \geq \delta) \to 0 \quad (\varepsilon \to 0). \tag{3.31}$$

4. Now by (3.11), (3.13)-(3.15), (3.20), (3.21), (3.23), (3.24), (3.28), (3.31), and Lemma 3.1 it follows that

$$\mathcal{L}(P_{n,[0,\rho]}^{(k)}) \xrightarrow{w} \mathcal{L}(P_{\infty,[0,\rho]}^{(k)}) \quad (n \to \infty) \tag{3.32}$$

for some random variable $P_{\infty,[0,\rho]}^{(k)}$. Eventually (3.12), (3.32), and Lemma 3.1 yield

$$\mathcal{L}(P_n^{(k)}) \xrightarrow{w} \mathcal{L}(P_\infty^{(k)}) \quad (n \to \infty)$$

for some $P_\infty^{(k)}$.

5. It remains to show that

$$\mathcal{L}(Q_n^{(k)}) \xrightarrow{w} \mathcal{L}(P_\infty^{(k)}) \quad (n \to \infty). \tag{3.33}$$

This can be done similarly as before. It follows from Lemma 3.2 and the law of large numbers that

$$\mathcal{L}(Q_{n,[\varepsilon,\rho]}^{(k)}) \xrightarrow{w} \mathcal{L}(P_{\infty,[\varepsilon,\rho]}^{(k)}) \quad (n \to \infty). \tag{3.34}$$

If we can show that $\{\mathcal{L}(Q_{n,t})\}_{n>1}$ is weakly convergent to $\operatorname{Exp} t(\gamma, 0, \eta)$ $(t \geq 0)$, then all of 2.-4. can be translated to this setting, so that we obtain some random variable $Q_\infty^{(k)}$ with

$$\mathcal{L}(Q_n^{(k)}) \xrightarrow{w} \mathcal{L}(Q_\infty^{(k)}) \quad (n \to \infty);$$

then it follows from (3.34) and Remark 3.2 that

$$\mathcal{L}(Q_\infty^{(k)}) = \mathcal{L}(P_\infty^{(k)}),$$

hence (3.33) holds. However, the convergence of $\{\mathcal{L}(Q_{n,t})\}_{n>1}$ can be established as follows:

Let $\mathcal{A}_n := (0, 0, \eta - \eta_{\lambda/n})$, $\mathcal{A} := (0, 0, \eta)$ ($\eta_{\lambda/n}$ as in (3.19)). Then ${}^\circ\mathcal{A}_n({}^\circ f) \to {}^\circ\mathcal{A}({}^\circ f)$ $(n \to \infty)$ $({}^\circ f \in C_b^\infty(I\!R^3))$, hence by Propositions 1.2, 1.3

$$\mathcal{L}(\overline{R}_{n,t}) \xrightarrow{w} \operatorname{Exp} t^\circ\mathcal{A} \quad (n \to \infty) \quad (t \geq 0) \tag{3.35}$$

and thus

$$\mathcal{L}(R_{n,t}) \xrightarrow{w} \varepsilon_{\circ\gamma t} * \operatorname{Exp} t^\circ\mathcal{A} \quad (n \to \infty) \quad (t \geq 0). \tag{3.36}$$

Since $\tilde{R}_{n,t} = R_{n,t} + \frac{1}{2}[\overline{R}_{n,t}, \frac{\gamma}{n}]$, it follows from (3.35) and (3.36) that

$$\mathcal{L}(\tilde{R}_{n,t}) \xrightarrow{w} \varepsilon_{\circ\gamma t} * \operatorname{Exp} t^\circ\mathcal{A} \quad (n \to \infty) \quad (t \geq 0),$$

and hence by Propositions 1.2, 1.3, and the identification $\mathcal{A} \leftrightarrow {}^\circ\mathcal{A}$

$$\mathcal{L}(Q_{n,t}) \xrightarrow{w} \operatorname{Exp}(\gamma, 0, \eta) \quad (n \to \infty) \quad (t \geq 0). \square$$

3.1.4 Tutubalin's theorem

In this section we present a limit theorem for i.i.d. $\mathbb{H}$-valued random variables with finite (euclidean) second moment due to Tutubalin (1964), which is in some sense special since the norming mappings are not endomorphisms of $(\mathbb{H},\cdot)$ but only of $(\mathbb{R}^3,+)$, the centering is performed with respect to "+" rather than "·", and also the limiting measure is a Gaussian distribution on $(\mathbb{R}^3,+)$.
For $a > 0$ define $\theta_a : \mathbb{R}^3 \to \mathbb{R}^3$ by

$$\theta_a(x', x'', x''') := (\sqrt{a}x', \sqrt{a}x'', a\sqrt{a}x''').$$

Theorem 3.5 *Let $\mu \in M^1(\mathbb{H})$ with*

$$\int_{\mathbb{H}} x'''^2 \mu(dx) < \infty,$$

$$b := \Big(\int_{\mathbb{H}} x'\mu(dx), \int_{\mathbb{H}} x''\mu(dx), 0\Big) \in \mathbb{H}\backslash\{0\},$$

$$\sigma_{a',a''} := \int_{\mathbb{R}^3} (a'^2 x'^2 + a''^2 x''^2)({}^\circ\mu * \varepsilon_{-{}^\circ b})(dx) \in]0,\infty[\quad ((a',a'') \in \mathbb{R}^2\backslash\{0\}).$$

Then there is a full Gaussian measure γ on $(\mathbb{R}^3,+)$ such that

$$\theta_{n^{-1}}({}^\circ(\mu^{*n}) * \varepsilon_{-n{}^\circ b}) \xrightarrow{w} \gamma \quad (n \to \infty).$$

Remark 3.3 *The additional assumptions on $\sigma_{a',a''}$ can be made since we do not have to consider the euclidean case.*

Proof of Theorem 3.5: Let $\{X_n\}_{n\ge 1}$ be i.i.d. $\mathbb{H}$-valued random variables distributed according to the law μ. Put $\overline{X_n} := {}^\circ X_n - {}^\circ b$. Then we get

$$\begin{aligned}\prod_{j=1}^n X_j &= \sum_{j=1}^n X_j + \frac{1}{2}\sum_{1\le i<j\le n} [\overline{X_i} + b, \overline{X_j} + b] \\ &= \sum_{j=1}^n X_j + \frac{1}{2}\sum_{1\le i<j\le n} [\overline{X_i}, \overline{X_j}] + D_n,\end{aligned}$$

$$\begin{aligned}D_n &= \frac{1}{2}\sum_{j=1}^n (n-j)[\overline{X_j}, b] + \frac{1}{2}\sum_{j=1}^n (j-1)[b, \overline{X_j}] \\ &= \frac{1}{2}\sum_{j=1}^n (n+1-2j)[\overline{X_j}, b].\end{aligned}$$

Clearly,

$$\mathrm{Var}\,(q(\sum_{j=1}^n X_j + \frac{1}{2}\sum_{1\le i<j\le n} [\overline{X_i}, \overline{X_j}])) = O(n^2) \quad (n \to \infty) \tag{3.37}$$

and by our assumption on $\sigma_{a',a''}$

$$\operatorname{Var}(q(D_n)) = \frac{1}{4}\sum_{j=1}^{n}(n+1-2j)^2 E(q([\overline{X_1},b])^2) = \Theta(n^3) \quad (n\to\infty). \tag{3.38}$$

By (3.37), (3.38), and the fact that $E[\overline{X_i},\overline{X_j}] = [E(\overline{X_i}),E(\overline{X_j})] = 0$ $(i\neq j)$, in the third component we only have to take into account the expression $q(D_n)$, i.e. it suffices to study the limit behavior of the $\mathbb{H}$-valued random variable

$$Z_n := \left(\frac{1}{\sqrt{n}}\sum_{j=1}^{n}\overline{X_j}', \frac{1}{\sqrt{n}}\sum_{j=1}^{n}\overline{X_j}'', \frac{1}{2n\sqrt{n}}\sum_{j=1}^{n}(n+1+2j)(\overline{X_j}'b''-\overline{X_j}''b')\right).$$

Let $\psi_n(u',u'',u''')$ resp. $\varphi(u',u'')$ denote the Fourier transform (on $(\mathbb{R}^3,+)$ resp. $(\mathbb{R}^2,+)$) of $\mathcal{L}(Z_n)$ resp. $\mathcal{L}(\overline{X_1}',\overline{X_1}'')$. We get (by the Taylor expansion of φ and the fact that $\mathcal{L}(\overline{X_1'},\overline{X_1''})$ is centered)

$$\begin{aligned}
\psi_n(u',u'',u''') &= \prod_{j=1}^{n}\varphi\left(\frac{u'}{\sqrt{n}}+u'''\frac{b''(n+1-2j)}{2n\sqrt{n}}, \frac{u''}{\sqrt{n}}-u'''\frac{b'(n+1-2j)}{2n\sqrt{n}}\right)\\
&= \prod_{j=1}^{n}\Big\{1+\frac{1}{2}\frac{\partial^2\varphi}{\partial u'^2}(0,0)\cdot\left(\frac{u'}{\sqrt{n}}+u'''\frac{b''(n+1-2j)}{2n\sqrt{n}}\right)^2\\
&\quad+\frac{1}{2}\frac{\partial^2\varphi}{\partial u''^2}(0,0)\cdot\left(\frac{u''}{\sqrt{n}}-u'''\frac{b'(n+1-2j)}{2n\sqrt{n}}\right)^2\\
&\quad+\frac{\partial^2\varphi}{\partial u'\partial u''}(0,0)\left(\frac{u'}{\sqrt{n}}+u'''\frac{b''(n+1-2j)}{2n\sqrt{n}}\right)\left(\frac{u''}{\sqrt{n}}-u'''\frac{b'(n+1-2j)}{2n\sqrt{n}}\right)\\
&\quad+o\left(\frac{1}{n}\right)\Big\}.
\end{aligned}$$

By our assumptions it follows that

$$\log\lim_{n\to\infty}\psi_n(u',u'',u''') = \frac{1}{2}E[(u'\overline{X_1}'+u''\overline{X_1}'')^2+\frac{u'''^2}{12}(b''\overline{X_1}'+b'\overline{X_1}'')^2],$$

which is, as follows easily by our assumptions, a non-degenerate quadratic form in $u=(u',u'',u''')$. □

Observe that the explicit form of the third coordinate X_n''' of the random variables X_n has no influence on the limit behavior.

Remark 3.4 *It seems that with the definition of $\mathbb{H}$ used in this text, Tutubalin's result and its proof can be formulated a little more elegantly than in the original paper of Tutubalin (1964), who uses the same kind of definition of $\mathbb{H}$ as Hulanicki (1976). The reason is that with us, "+" is really the addition operation on the Lie algebra of $\mathbb{H}$.*

3.1.5 Triangular systems

A classical theorem in probability theory says that every limit of an infinitesimal triangular system of probability measures on $\mathbb{R}$ is infinitely divisible. Actually this turned

out to be true for very general abelian groups, such as e.g. all Banach spaces and all divisible locally compact abelian groups (see e.g. Ruzsa (1988)). Gangolli (1964) generalized the theorem to symmetric spaces, and Carnal (1966) resp. Neuenschwander (1995g), (1995f) proved it for compact groups resp. euclidean motion groups and discrete subgroups of simply connected nilpotent Lie groups under the additional proviso that the measures commute within each row of the triangular system. The theorem is also true on hermitean hypergroups (cf. Bloom, Heyer (1995), Proposition 5.3.11). There are also other treatments available in the literature on triangular systems in several contexts, however the following assertion seems to be new. Here for $\mathbb{H}$, the main point will be (as in Neuenschwander (1995f)) to verify the two conditions of Wehn (1962) and Siebert (1981), which allow us to approximate the row products of the triangular system by certain Poisson laws similarly as in the classical situation: They follow by imposing some "local centering" condition and by considering the system on the underlying euclidean space $(\mathbb{R}^3, +)$.
The first part of this section is essentially based on Neuenschwander (1992); however, the result is formulated (for $\mathbb{H}$) in a more general setting.

Definition 3.1 *A [locally centered commutative infinitesimal] triangular system ([l.c.c.i.]t.s.) on the simply connected nilpotent Lie group G is a double array $\Delta = \{\mu_{n,j}\}_{n\geq 1;1\leq j\leq k(n)} \subset M^1(G)$ satisfying*

$$[\mu_{n,i} * \mu_{n,j} = \mu_{n,j} * \mu_{n,i} \quad (n \geq 1 ; 1 \leq i, j \leq k(n)) \quad (\textit{commutativity}).$$

$$\min_{1\leq j\leq k(n)} \mu_{n,j}(U) \to 1 \quad (n \to \infty)$$

$$(U \text{ a Borel neighborhood of } 0) \quad (\textit{infinitesimality}).$$

$$\int_{\|x\|<\rho} x\mu_{n,j}(dx) = 0 \quad \textit{for some } \rho > 0$$

$$(n \geq 1; 1 \leq j \leq k(n)) \quad (\textit{local centration}).]$$

The array Δ is said to be relatively compact or convergent to $\mu \in M^1(G)$, if the sequence of convolution products $\{\mu_{n,1} * \mu_{n,2} * \ldots * \mu_{n,k(n)}\}_{n\geq 1}$ has the respective property (with respect to the weak topology). In the euclidean case, commutativity holds automatically, and we will speak merely of a *[locally centered infinitesimal] triangular system* ([l.c.i.]t.s.) in this case.
The following lemma is similar to Lemma 2.4:

Lemma 3.3 *If $\{\mu_{n,j}\}_{n\geq 1;1\leq j\leq k(n)}$ is a convergent c.t.s. on $\mathbb{H}$, then the t.s. $\{{}^{\circ}\mu_{n,j}\}_{n\geq 1;1\leq j\leq k(n)}$ on $(\mathbb{R}^3, +)$ is relatively compact.*

Proof: Consider $\mathbb{H}$-valued random variables $\{X_{n,j}\}_{n\geq 1;1\leq j\leq k(n)}$ with distribution $\mathcal{L}(X_{n,j}) = \mu_{n,j}$ such that $X_{n,1}, X_{n,2}, \ldots, X_{n,k(n)}$ are independent for all $n \geq 1$. Since the sequences $\{\mathcal{L}(X_n)\}_{n\geq 1}$ and $\{\mathcal{L}(\overline{X_n})\}_{n\geq 1}$, where

$$\begin{aligned} X_n &:= X_{n,1} \cdot X_{n,2} \cdot \ldots \cdot X_{n,k(n)}, \\ \overline{X_n} &:= X_{n,k(n)} \cdot X_{n,k(n)-1} \cdot \ldots \cdot X_{n,1}, \end{aligned}$$

are both uniformly tight, the same is true for $\{\mathcal{L}(X_n + \overline{X_n})\}_{n>1}$. Now the same calculation as in (2.16) yields

$$X_n + \overline{X_n} = 2\sum_{j=1}^{k(n)} X_{n,j}$$

and the assertion follows from Prohorov's theorem.

Definition 3.2 *The accompanying system $\Delta^{\sim} = \{\nu_{n,j}\}_{n>1;1\leq j\leq k(n)}$ of the c.i.t.s. $\{\mu_{n,j}\}_{n>1;1<j\leq k(n)}$ on the simply connected nilpotent Lie group G is given by the Poisson measures*

$$\tilde{\mu}_{n,j} := \exp(\mu_{n,j} - \varepsilon_0) \quad (n \geq 1; 1 \leq j \leq k(n)).$$

It is not difficult to verify that $\Delta^{\sim}$ is a c.i.t.s., too (see also Siebert (1981), p.146). If $G = (\mathbb{R}^d, +)$, then by the so-called "accompanying laws theorem", a l.c.i.t.s. Δ converges iff $\Delta^{\sim}$ converges, and in this case the limits of Δ and $\Delta^{\sim}$ coincide (cf. Rvačeva (1962), Theorem 2.1). (Note that Definition 3.2 is in general not coincident with the definition of the accompanying system normally used in the classical case, since we have, by the local centration, no additional shifts). For groups, conditions for the validity of the approximation of a c.i.t.s. by the system of the accompanying laws were given by Wehn (1962) and Siebert (1981). It is now an immediate consequence of Lemma 3.3 and the limit behavior of convergent i.t.s. on $(\mathbb{R}^3, +)$ that for convergent l.c.c.i.t.s. on $\mathbb{H}$ Wehn's conditions are always fulfilled. Eventually, the strong root compactness yields the assertion of infinite divisibility.

Theorem 3.6 *Assume $\Delta = \{\mu_{n,j}\}_{n\geq 1;1\leq j<k(n)}$ is a convergent l.c.c.i.t.s. on $\mathbb{H}$. Then $\Delta^{\sim}$ is convergent with the same limit, which is then infinitely divisible on $\mathbb{H}$ and thus embeddable into a c.c.s. on $\mathbb{H}$.*

Proof: This is an application of Remarks 1 and 4 on pp.148ff. in Siebert (1981). Let $\{x_1, x_2, x_3\} \subset C^{\infty}(\mathbb{H})$ be a system of canonical coordinates with compact support such that $x_i(-x) = -x_i(x)$ $(i = 1, 2, 3; x \in \mathbb{H})$. Let Φ be a so-called *Hunt function*, i.e. $\Phi \in C^{\infty}(\mathbb{H}), \Phi(x) = \Phi(-x) \geq 0$ $(x \in \mathbb{H}), \Phi(x) = x_1^2 + x_2^2 + x_3^2$ $(x \in U_0)$ for some Borel neighborhood U_0, and Φ is bounded away from 0 on $\mathrm{cpl}\, U$ for any Borel neighborhood U of 0 (cf. Siebert (1981), p.128). Then by Siebert (1981), Remarks 1 and 4 on pp.148ff., $\Delta^{\sim}$ converges to the same limit $\mu \in M^1(\mathbb{H})$ as Δ if the following conditions due to Wehn (1962) hold:

(W1) $\limsup_{n\to\infty} \sum_{j=1}^{k(n)} \int_{\mathbb{H}} \Phi(x)\mu_{n,j}(dx) < \infty$.

(W2) $\limsup_{n\to\infty} \sum_{j=1}^{k(n)} |\int_{\mathbb{H}} x_i(x)\mu_{n,j}(dx)| < \infty \quad (i = 1, 2, 3)$.

The l.c.i.t.s. $\{{}^{\circ}\mu_{n,j}\}_{n\geq 1;1\leq j\leq k(n)}$ is relatively compact by Lemma 3.3, and it follows from the local centration and Rvačeva (1962), Lemma 2.5 that the sums

$$\sum_{j=1}^{k(n)} \int_{\mathbb{R}^3} \frac{||x||^2}{1 + ||x||^2}\, {}^{\circ}\mu_{n,j}(dx) \tag{3.39}$$

are bounded, so (W1) holds. Now (W2) follows from (W1) and the local centering condition. By the (strong) root compactness of $\mathbb{H}$, the set of infinitely divisible probability measures on $\mathbb{H}$ is weakly closed (Heyer (1977), Theorem 3.1.28). So in our situation μ, being the limit of Poisson laws, is indeed infinitely divisible and thus embeddable into a c.c.s. on $\mathbb{H}$ by Corollary 1.2. □

Similarly as in the i.i.d. case for general simply connected nilpotent Lie groups (cf. Corollary 1.3 and the following remark), one can formulate the following "transfer principle", saying that limit theorems for l.c.i.t.s. on $(\mathbb{R}^3, +)$ whose measures within each row commute on $\mathbb{H}$ have a canonical counterpart on $(\mathbb{H}, \cdot)$:

For $\Delta = \{\mu_{n,j}\}_{n\geq 1; 1\leq j\leq k(n)}$, consider ${}^\circ\Delta = \{{}^\circ\mu_{n,j}\}_{n\geq 1; 1\leq j\leq k(n)}$.

Theorem 3.7 *Let Δ be a l.c.c.i.t.s. on $(\mathbb{H}, \cdot)$. Then, if ${}^\circ\Delta$ converges to* $\operatorname{Exp}{}^\circ\mathcal{A}$, *it follows that Δ converges to* $\operatorname{Exp}\mathcal{A}$.

Proof: By the proof of Theorem 3.6 (applied to $(\mathbb{R}^3, +)$) it follows that $({}^\circ\Delta)^\sim$ converges to $\operatorname{Exp}{}^\circ\mathcal{A}$. By Propositions 1.2, 1.3, and the identification $\mathcal{A} \leftrightarrow {}^\circ\mathcal{A}$ we get that $\Delta^\sim$ converges to $\operatorname{Exp}\mathcal{A}$. Now the proof of Theorem 3.6 and (the other direction of) Siebert (1981), Remark 1 on p.148 (applied to $(\mathbb{H}, \cdot)$) yield the assertion. □

The "transfer principle" of Theorem 3.7 also holds the other way round if the limit measure is Gaussian on $\mathbb{H}$ by Theorem 2.1. As in Corollary 1.3, what is missing in order to prove this assertion in general is the full uniqueness property of c.c.s. on $(\mathbb{H}, \cdot)$.

Now we turn to non-commutative t.s. of symmetric measures on $\mathbb{H}$. Up to now, limit theorems for non-commutative t.s. only have asserted the embeddability of the limit in a c.c.h. (cf. Feinsilver(1978), Siebert (1982), Heyer, Pap (1996), Pap (1996a, 1996b)), but not necessarily in a c.c.s. Theorem 3.8 seems to be the first general result concerning embeddability of limits of non-commutative t.s. in a c.c.s.

If the t.s. on $\mathbb{H}$ consists of symmetric measures, then the analogue of Lemma 3.3 holds (since in the proof of Lemma 3.3 commutativity is only used to reverse the order of the $\mu_{n,1}, \mu_{n,2}, \ldots, \mu_{n,k(n)}$):

Lemma 3.4 *If $\{\mu_{n,j}\}_{n>1; 1<j\leq k(n)}$ is a convergent t.s. on $\mathbb{H}$ and all $\mu_{n,j}$ are symmetric, then the t.s. $\{{}^\circ\mu_{n,j}\}_{n>1; 1<j\leq k(n)}$ on $(\mathbb{R}^3, +)$ is relatively compact.*

Proof: The proof is parallel to that of Lemma 3.3 as soon as the uniform tightness of $\{\mathcal{L}(\overline{X}_n)\}_{n\geq 1}$ is proved. This, however, can be shown as follows: Since $\{\mathcal{L}(X_n)\}_{n>1}$ is uniformly tight, the same holds for $\{\mathcal{L}(-X_n)\}_{n>1}$. But by the symmetry

$$\begin{aligned}
\mathcal{L}(-X_n) &= \mathcal{L}(-(X_{n,1} \cdot X_{n,2} \cdot \ldots \cdot X_{n,k(n)})) \\
&= \mathcal{L}((-X_{n,k(n)}) \cdot (-X_{n,k(n)-1}) \cdot \ldots \cdot (-X_{n,1})) \\
&= \mathcal{L}(-X_{n,k(n)}) * \mathcal{L}(-X_{n,k(n)-1}) * \ldots * \mathcal{L}(-X_{n,1}) \\
&= \mathcal{L}(X_{n,k(n)}) * \mathcal{L}(X_{n,k(n)-1}) * \ldots * \mathcal{L}(X_{n,1}) \\
&= \mathcal{L}(\overline{X}_n). \square
\end{aligned}$$

Furthermore, we will have to use the following technical fact, which was established by Carnal (1966) (see also Heyer (1977), Corollary 6.6.3):

Lemma 3.5 *Let $\{x_1, x_2, \ldots, x_m\} \subset \mathbb{R}^\ell, \varepsilon > 0, \|x_i\| < \varepsilon$ $(1 \leq i \leq m), p \in \mathbb{N}, p \leq m$. Then the set $\{1, 2, \ldots, m\}$ can be decomposed into p disjoint classes $\Pi_1, \Pi_2, \ldots, \Pi_p$ such* *that for some constant $C(\ell,p) > 0$ one has*

$$\|\sum_{i\in \Pi_r} x_i - \sum_{i\in \Pi_s} x_i\| < C(\ell,p)\varepsilon \quad (1 \le r, s \le p).$$

Theorem 3.8 *Assume $\{\mu_{n,j}\}_{n\ge 1, 1<j\le k(n)}$ is an i.t.s. on $\mathbb{H}$ which converges to $\mu \in M^1(\mathbb{H})$. If all μ_{nj} are symmetric, then μ is embeddable into a c.c.s. on $\mathbb{H}$ (and thus infinitely divisible).*

Proof: As in the proof of Theorem 3.6, it follows from Lemma 3.4, the symmetry, and Rvačeva (1962), Lemma 2.5 that also here the sums (3.39) are bounded, so condition (S') (and thus (S)) of Siebert (1982), Lemma 3.1 holds, which yields the applicability of his Theorem 3.6. Since $\tilde{C}_2(\mathbb{H})$ is separable, we may choose a sequence $\{f_n\}_{n\ge 1}$ which is dense in $\tilde{C}_2(\mathbb{H})$. By Lemma 3.5, applied to vectors of the form

$$\{\int_{\mathbb{H}} |f_i(x) - f_i(0)|\mu_{n,j}(dx)\}_{1\le i\le \ell}$$

(by the infinitesimality and Rvačeva (1962), Lemma 2.2), Lemma 3.4, a diagonal argument, and Siebert (1982), Theorems 3.6, 5.7, 4.3 and its proof (statement "$dB_t f/dt = \mathcal{A}_t f$" on last line on p. 381) we get that the limiting Lipschitz continuous c.c.h. has a generating family $\{\mathcal{A}_t\}_{t\ge 0} = \{\mathcal{A}\}$, hence comes from a c.c.s. □

3.2 A.s. theorems

3.2.1 The Marcinkiewicz-Zygmund law of large numbers

Several extensions of the classical strong law of large numbers are known in the literature (see e.g. Furstenberg (1963), Tutubalin (1969), Guivarc'h (1976)). For simply connected nilpotent Lie groups (e.g.), the analogue of the classical Kolmogorov strong law of large numbers follows e.g. from Guivarc'h (1976):

Theorem 3.9 *Let $G \simeq \mathbb{R}^d$ be a simply connected nilpotent Lie group, $K :=]-1,1[^d$, and*

$$|x|_K := \inf\{n \ge 1 : x \in K^n\}.$$

Assume $\mu \in M^1(G)$ with

$$\int_G |x|_K \mu(dx) < \infty$$

and

$$\int_G h(x)\mu(dx) = 0$$

for every homomorphism $h : G \to \mathbb{R}$. Let $\{X_n\}_{n\ge 1}$ be i.i.d. G-valued random variables obeying to the law μ. Then

$$|\prod_{j=1}^{n} X_j|_K / n \xrightarrow{a.s.} 0 \quad (n \to \infty).$$

The aim of this section is to prove the following analogue on $\mathbb{H}$ of the classical Marcinkiewicz-Zygmund strong law of large numbers (cf. Chow, Teicher (1978), Theorem 5.2.2, Neuenschwander (1995b)). For a version for the euclidean motion groups and the diamond groups, see Neuenschwander (1995h).

Theorem 3.10 *Let $|.|$ be an arbitrary homogeneous norm on $\mathbb{H}$. Assume $X_1, X_2, \ldots$ are i.i.d. $\mathbb{H}$-valued random variables. Then for any $p \in]0, 2[$*

$$\delta_{n^{-1/p}}\Big(\prod_{j=1}^{n}(X_j \cdot c)\Big) \xrightarrow{a.s.} 0 \quad (n \to \infty) \tag{3.40}$$

for some $c \in \mathbb{H}$ iff $E|X_1|^p < \infty$. If so, then in case $1 \le p < 2$

$$(c', c'') = (-E(X_1)', -E(X_1)''),$$

while c''' and, in case $0 < p < 1$, also (c', c'') can be chosen arbitrarily.

A sharpened version (Baum-Katz theorem) will be given in 3.2.2.
By the equivalence of all homogeneous norms it follows that $||(x', x'')||, \sqrt{|x'''|} \le A|x|$ for some $A > 0$, which we will use implicitly in the following.
For the proof of Theorem 3.10 in the case $1 \le p < 2$ we need the following lemma, which is similar to Kronecker's lemma (cf. Chow, Teicher (1978), Lemma 5.1.2).

Lemma 3.6 *For any sequences $\{a_n\}_{n\ge1} \subset \mathbb{R}^d$, $\{b_n\}_{n\ge1} \subset]0, \infty[$ such that $b_{n+1} \ge b_n$ $(n \ge 1)$, $b_n \to \infty$ $(n \to \infty)$, and*

$$\frac{1}{b_n}\sum_{j=1}^{n}\frac{a_j}{b_j} \to 0 \quad (n \to \infty)$$

we have

$$\frac{1}{b_n^2}\sum_{j=1}^{n} a_j \to 0 \quad (n \to \infty).$$

Proof: For $\varepsilon > 0$, choose $N \in \mathbb{N}$ such that

$$\frac{1}{b_n}||\sum_{j=1}^{n}\frac{a_j}{b_j}|| \le \frac{\varepsilon}{3} \quad (n \ge N).$$

Then by summation by parts we get

$$\begin{aligned}
\frac{1}{b_n^2}||\sum_{j=1}^{n} a_j|| &= \frac{1}{b_n^2}||\sum_{j=1}^{n} b_j\frac{a_j}{b_j}|| \\
&\le \frac{1}{b_n}||\sum_{j=1}^{n}\frac{a_j}{b_j}|| + \frac{1}{b_n^2}||\sum_{j=1}^{n-1}(b_{j+1} - b_j)\sum_{i=1}^{j}\frac{a_i}{b_i}|| \\
&\le \frac{\varepsilon}{3} + \frac{1}{b_n^2}||\sum_{j=1}^{N-1}(b_{j+1} - b_j)\sum_{i=1}^{j}\frac{a_i}{b_i}|| + \frac{\varepsilon}{3} \\
&\le \varepsilon
\end{aligned}$$

for n large enough.□

Proof of Theorem 3.10: 1. The "only if"-part may be proved similarly as in the classical situation (cf. Chow, Teicher (1978), p.122): Since

$$\begin{aligned}
\delta_{n^{-1/p}}(X_n \cdot c) &= \delta_{n^{-1/p}}((-\prod_{j=1}^{n-1}(X_j \cdot c)) \cdot \prod_{j=1}^{n}(X_j \cdot c)) \\
&= \delta_{(\frac{n}{n-1})^{-1/p}}(\delta_{(n-1)^{-1/p}}(-\prod_{j=1}^{n-1}(X_j \cdot c))) \cdot \delta_{n^{-1/p}}(\prod_{j=1}^{n}(X_j \cdot c)) \\
&\overset{a.s.}{\to} 0 \quad (n \to \infty)
\end{aligned}$$

by (3.40), it follows from the Borel-Cantelli Lemma that

$$\sum_{n=1}^{\infty} P(|X_1 \cdot c| > n^{1/p}) < \infty,$$

which implies

$$E|X_1 \cdot c|^p < \infty$$

by Chow, Teicher (1978), Corollary 4.1.3. So

$$\begin{aligned}
E|X_1|^p &= E|X_1 \cdot c \cdot (-c)|^p \\
&\leq CE(|X_1 \cdot c| + |-c|)^p \\
&< \infty.
\end{aligned}$$

2. Assume $0 < p < 1$, $E|X_1|^p < \infty$, and let $c \in I\!H$ be arbitrary. By considering $X_j \cdot c$ instead of X_j we may w.l.o.g. assume $c = 0$. By the classical Marcinkiewicz-Zygmund strong law of large numbers we have

$$(n^{-1/p}(\sum_{j=1}^{n} X_j)', n^{-1/p}(\sum_{j=1}^{n} X_j)'') \overset{a.s.}{\to} 0 \quad (n \to \infty). \tag{3.41}$$

so we have to prove

$$n^{-2/p}(\prod_{j=1}^{n} X_j)''' \overset{a.s.}{\to} 0 \quad (n \to \infty). \tag{3.42}$$

This is equivalent to

$$n^{-2/p}(\sum_{j=1}^{n} X_j)''' + \frac{1}{2} n^{-2/p}(\sum_{1 \leq i < j \leq n} [X_i, X_j])''' \overset{a.s.}{\to} 0 \quad (n \to \infty).$$

The first summand tends to 0 a.s. by the classical Marcinkiewicz-Zygmund strong law of large numbers. So it remains to show that

$$n^{-2/p}(\sum_{1 \leq i < j \leq n} [X_i, X_j])''' \overset{a.s.}{\to} 0 \quad (n \to \infty). \tag{3.43}$$

But for this we have

$$|n^{-2/p}(\sum_{1\le i<j\le n}[X_i, X_j])'''| = O((n^{-1/p}\sum_{j=1}^{n}\|(X_j', X_j'')\|)^2)$$
$$\xrightarrow{a.s.} 0 \quad (n \to \infty)$$

by the classical Marcinkiewicz-Zygmund strong law of large numbers, which proves (3.43) and thus (3.42).

3. Assume $1 \le p < 2$, $E|X_1|^p < \infty$, and again w.l.o.g. $c = 0$, $E(X_1)' = E(X_1)'' = 0$, and let $c''' \in \mathbb{R}$ be arbitrary. Hence, as above, in order to prove (3.40), it remains to show (3.43). Define the (real-valued) random variables

$$Z_n := [n^{-1/p}\sum_{j=1}^{n-1} X_j, X_n]''',$$
$$\overline{Z_n} := Z_n \cdot 1\{n^{-1/p}\|((\sum_{j=1}^{n-1} X_j)', (\sum_{j=1}^{n-1} X_j)'')\| \le 1\}.$$

W.l.o.g. we may assume that there is a $\mathbb{H}$-valued random variable X which is distributed like X_1 and independent of $\{X_n\}_{n\ge 1}$. We have

$$E\|(X', X'')\| < \infty,$$

so since a.s.

$$\begin{aligned} &P(|\overline{Z_n}| \ge x | X_1, X_2, \ldots, X_{n-1}) \\ &\le P(\|(X_n', X_n'')\| \ge x | X_1, X_2, \ldots, X_{n-1}) \\ &= P(\|(X', X'')\| \ge x | X_1, X_2, \ldots, X_{n-1}) \quad (0 \le x < \infty), \end{aligned}$$

the Theorem in Chatterji (1969) yields

$$n^{-1/p}\sum_{j=1}^{n}(\overline{Z_j} - \alpha_j) \xrightarrow{a.s.} 0 \quad (n \to \infty),$$

where

$$\alpha_n := E(\overline{Z_n}|\overline{Z_1}, \overline{Z_2}, \ldots, \overline{Z_{n-1}}).$$

By the classical Marcinkiewicz-Zygmund strong law of large numbers, there is a.s. an N (random) such that a.s.

$$Z_n = \overline{Z_n} \quad (n \ge N) \tag{3.44}$$

and

$$E(\overline{Z_n}|X_1, X_2, \ldots, X_{n-1}) = E(Z_n|X_1, X_2, \ldots, X_{n-1})$$
$$= 0 \quad (n \ge N)$$

(since $E(X_1)' = E(X_1)'' = 0$), hence a.s.

$$\alpha_n = 0 \quad (n \ge N). \tag{3.45}$$

Thus by (3.44),(3.45)

$$n^{-1/p} \sum_{j=1}^{n} Z_j \xrightarrow{a.s.} 0 \quad (n \to \infty).$$

By Lemma 3.6 (with

$$b_n := n^{1/p},$$

$$a_n := b_n Z_n - |\sum_{j=1}^{n-1} X_j, X_n|''')$$

this proves (3.43).

3. The fact that (c', c'') is uniquely determined in case $1 \le p < 2$ follows from (3.41).□

Now we apply the subsequence principle of 2.3.3 to the Marcinkiewicz-Zygmund strong law of large numbers.

Corollary 3.1 *Assume $\{X_n\}_{n \ge 1}$ is a uniformly tight sequence of $\mathbb{H}$-valued random variables such that $\sup_{n \ge 1} E|X_n|^p < \infty$ $(1 \le p < 2)$, where $|.|$ is any homogeneous norm on $\mathbb{H}$. Then there exists a $\mathbb{H}$-valued random variable α and a subsequence $\{n'\} \subset \{n\}$ such that for any subsequence $\{n(m)\}_{m \ge 1} \subset \{n'\}$ we have*

$$\delta_{n^{-1/p}}(\prod_{j=1}^{m} (X_{n(j)} \cdot \alpha)) \xrightarrow{a.s.} 0 \quad (m \to \infty).$$

Proof: The proof is analogous to that of Corollary 2.4 using Theorem 3.10, the classical Marcinkiewicz-Zygmund strong law of large numbers, and Lemma 2.20 for $\delta_{n^{-1/p}}(\ldots)$. Here, the limit statute is

$$A = A_1 \cup A_2,$$

$$A_1 := \{(\nu, x) \in M^1(\mathbb{H}) \times \mathbb{H}^{\infty} : \delta_{n^{-1/p}}(\prod_{j=1}^{n} (x_j \cdot \alpha(\nu))) \to 0 \quad (n \to \infty),$$

$$\lim_{n \to \infty} \sum_{j=1}^{n} ||\delta_{n^{-1/p}}(x_j)|| < \infty, \int_{\mathbb{H}} |\xi|^p \nu(d\xi) < \infty\},$$

$$A_2 := \{(\nu, x) \in M^1(\mathbb{H}) \times \mathbb{H}^{\infty} : \int_{\mathbb{H}} |\xi|^p \nu(d\xi) = \infty\},$$

where

$$p(\alpha(\nu)) := \int_{\mathbb{H}} p(x) \nu(dx),$$

$$q(\alpha(\nu)) := 0. \square$$

In a similar way (here $y = 0$ in Lemma 2.20) we get for $0 < p < 1$:

Corollary 3.2 *Assume $\{X_n\}_{n \ge 1}$ is a uniformly tight sequence of $\mathbb{H}$-valued random variables such that $\sup_{n \ge 1} E|X_n|^p < \infty$ $(0 < p < 1)$, where $|.|$ is any homogeneous norm on $\mathbb{H}$. Then there exists a subsequence $\{n'\} \subset \{n\}$ such that for any subsequence $\{n(m)\}_{m \ge 1} \subset \{n'\}$ we have*

$$\delta_{n^{-1/p}}(\prod_{j=1}^{m} X_{n(j)}) \xrightarrow{a.s.} 0 \quad (m \to \infty).$$

3.2.2 Rates of convergence in laws of large numbers

In this section, we prove two theorems giving preciser information on the rate of convergence in the Kolmogorov resp. Marcinkiewicz-Zygmund law of large numbers on $\mathbb{H}$. For $\mathbb{R}$, it was shown by Hsu-Robbins-Erdös that in the law of large numbers, complete convergence is equivalent to the finiteness of the second moment. Baum and Katz strengthened the Marcinkiewicz-Zygmund law of large numbers in the sense that there is not only strong convergence, but convergence of certain series which implies complete convergence.
The aim of this section is to carry over the theorems of Hsu-Robbins-Erdös resp. Baum-Katz to $\mathbb{H}$ (cf. Neuenschwander (1995b)).
A sequence $\{X_n\}_{n\geq 1}$ of random variables is said to *converge completely* to the random variable X if for every $\epsilon > 0$

$$\sum_{n=1}^{\infty} P(|X_n - X| > \epsilon) < \infty.$$

By the Borel-Cantelli Lemma, complete convergence implies a.s. convergence.
We will use the following consequence of the Hölder inequality:

Lemma 3.7 *Assume $x_1, y_1, x_2, y_2, \ldots, x_n, y_n \in \mathbb{H}$. Then*

$$\|\sum_{i=1}^{n}[x_i, y_i]\| \leq \max_{1\leq i\leq n} \|(x_i', x_i'')\| \cdot \sum_{i=1}^{n} \|(y_i', y_i'')\|.$$

Proof: By Hölder's Inequality,

$$\|\sum_{i=1}^{n}[x_i, y_i]\| \leq \Big(\sum_{i=1}^{n} \|(x_i', x_i'')\|^p\Big)^{1/p}\Big(\sum_{i=1}^{n} \|(y_i', y_i'')\|^q\Big)^{1/q}$$

for $p, q > 1, \frac{1}{p} + \frac{1}{q} = 1$. Now $q \to 1$ yields the assertion.□
First we carry over the theorem of Hsu-Robbins-Erdös (cf. Chow, Teicher (1978), Corollary 10.4.2):

Theorem 3.11 *Let $|.|$ a homogeneous norm on $\mathbb{H}$, and assume $\{X_n\}_{n>1}$ are i.i.d. $\mathbb{H}$-valued random variables. Then*

$$\sum_{n=1}^{\infty} P(|\delta_{n^{-1}}(\prod_{i=1}^{n}(X_i \cdot c))| \geq \rho) < \infty \quad \text{for every } \rho > 0 \tag{3.46}$$

iff

$$E|X_1|^2 < \infty, (c', c'') = (-E(X_1)', -E(X_1)'').$$

Proof: We first prove the "if"-direction: As in the proof of Theorem 3.10, we may assume w.l.o.g. $c = 0, E(X_1)' = E(X_1)'' = 0$. By Lemma 3.7,

$$|\delta_{n^{-1}}(\prod_{j=1}^{n} X_j)|$$

$$
\begin{aligned}
&\le n^{-1}\|\sum_{j=1}^{n}(X_j', X_j'')\| + \{n^{-2}|\sum_{j=1}^{n} X_j'''|\}^{1/2} \\
&\quad + (1/\sqrt{2})\{n^{-1}\max_{1<j<n}\|(X_j', X_j'')\| \cdot n^{-1}\sum_{j=1}^{n}\|(X_j', X_j'')\|\}^{1/2} \\
&= T_n^{(1)} + \sqrt{T_n^{(2)}} + (1/\sqrt{2})\sqrt{T_n^{(3)}\cdot T_n^{(4)}}.
\end{aligned}
$$

Suppose $\sigma > 0$. We have

$$\sum_{n=1}^{\infty} P(T_n^{(1)} \ge \sigma) < \infty \tag{3.47}$$

by Chow, Teicher (1978), Corollary 10.4.2 and

$$\sum_{n=1}^{\infty} P(T_n^{(2)} \ge \sigma) < \infty \tag{3.48}$$

by Chow, Teicher (1978), Theorem 10.4.1 (with $\alpha = 2, p = \gamma = 1$; it is easy to see that the theorem is also valid in case $EX \neq 0, \alpha > 1$). For $T_n^{(3)}$ we get

$$
\begin{aligned}
\sum_{n=1}^{\infty} P(T_n^{(3)} \ge \sigma) &\le \sum_{n=1}^{\infty} nP(\|(X_1', X_1'')\| \ge \sigma n) \\
&\le 1 + \int_1^{\infty}(t+1)P(\|(X_1', X_1'')\|^2 \ge \sigma^2 t^2)dt \\
&\le 1 + C\int_1^{\infty} P(\|(X_1', X_1'')\|^2 \ge s)ds \\
&\le 1 + CE\|(X_1', X_1'')\|^2 \\
&< \infty.
\end{aligned} \tag{3.49}
$$

Again by Chow, Teicher (1978), Corollary 10.4.2

$$\sum_{n=1}^{\infty} P(T_n^{(4)} - E\|(X_1', X_1'')\| \ge \sigma) < \infty. \tag{3.50}$$

Now w.l.o.g. $E\|(X_1', X_1'')\| > 0$, for otherwise (3.48) proves the "if"-direction. Since

$$
\begin{aligned}
&P(|\delta_{n^{-1}}(\prod_{i=1}^{n} X_i)| \ge \rho) \\
&\le P(T_n^{(1)} \ge \frac{\rho}{3}) + P(T_n^{(2)} \ge (\frac{\rho}{3})^2) \\
&\quad + P(T_n^{(3)} \ge (\frac{\rho}{3})^2(E\|(X_1', X_1'')\|)^{-1}) \\
&\quad + P(T_n^{(4)} - E\|(X_1', X_1'')\| \ge E\|(X_1', X_1'')\|),
\end{aligned}
$$

(3.47)-(3.50) yield the assertion.
As far as the "only if"-part is concerned, first observe that by (3.46) and the Borel-Cantelli Lemma

$$\frac{1}{n}\sum_{i=1}^{n}(X_i' + c', X_i'' + c'') \xrightarrow{a.s.} 0,$$

hence by the classical Marcinkiewicz-Zygmund law of large numbers it follows that $(E(X_1)', E(X_1)'') = (-c', -c'')$.

Now we show that $E|X_1|^2 < \infty$. For this, it suffices to prove, by Chow, Teicher (1978), Corollary 10.4.2, that $E|X_1'''| < \infty$. Observe that as in (2.16)

$$\prod_{i=1}^{n}(X_i \cdot c) + \prod_{i=1}^{n}(X_{n+1-i} \cdot c) = 2\sum_{i=1}^{n}(X_i \cdot c). \tag{3.51}$$

Now we may proceed similarly as in the proof of Chow, Teicher (1978), Corollary 10.4.2: Let $\{\tilde{X}_n\}_{n\geq 1}$ be an independent copy of the process $\{X_n\}_{n\geq 1}$, and put $Y_n = (X_n \cdot c) - (\tilde{X}_n \cdot c)$. Then we get by (3.51), the symmetry of Y_n, and Lévy's inequality (cf. Chow, Teicher (1978), Lemma 3.3.5)

$$\begin{aligned}
1 - P^n(|Y_1'''| < \beta) &= P(\max_{1\leq i\leq n} |Y_i'''| \geq \beta) \\
&\leq P(\max_{1\leq i\leq n} |(\sum_{j=1}^{i} Y_j)'''| \geq \frac{\beta}{2}) \\
&\leq 2hP(|(\sum_{i=1}^{n} Y_i)'''| \geq \frac{\beta}{2h}) \\
&\leq 4hP(|(\sum_{i=1}^{n}(X_i \cdot c))'''| \geq \frac{\beta}{4h}) \\
&\leq 8hP(|(\prod_{i=1}^{n}(X_i \cdot c))'''| \geq \frac{\beta}{4h}) \\
&\leq 8hP(|\prod_{i=1}^{n}(X_i \cdot c)| \geq H\sqrt{\frac{\beta}{4h}})
\end{aligned} \tag{3.52}$$

for some constant $H > 0$, hence for

$$\gamma = \frac{H}{\sqrt{4h}}$$

we get by (3.46) and (3.52)

$$\begin{aligned}
\infty &> \sum_{n=1}^{\infty}(1 - P^n(|Y_1'''| < n^2)) \\
&= \sum_{n=1}^{\infty} P(|Y_1'''| \geq n^2) \sum_{j=0}^{n-1} P^j(|Y_1'''| < n^2) \\
&\geq \sum_{n=1}^{\infty} nP(|Y_1'''| \geq n^2)[\frac{1}{n}\sum_{j=0}^{n-1}(1 - 8hP(|\prod_{i=1}^{j}(X_i \cdot c)| \geq \gamma j))].
\end{aligned}$$

The expression $[\ldots]$ tends to 1 as $n \to \infty$ by (3.46), and

$$\sum_{n=1}^{\infty} nP(|Y_1'''| \geq n^2) > \int_1^{\infty} tP(|Y_1'''| \geq (t+1)^2)dt$$

$$\geq \frac{1}{4}\int_1^{\infty} P(|Y_1'''| \geq s)ds$$

$$\geq \frac{1}{4}(E|Y_1'''| - 1).$$

so $E|Y_1'''| < \infty$ and thus, by Chow, Teicher (1978), Lemma 10.1.1, $E|(X_1 \cdot c)'''| < \infty$. Since $E(X_1)' = E(X_1)'' = 0$ it follows that

$$E(X_1 \cdot c)''' = E(X_1)''' + c'''.$$

so $E|X_1'''| < \infty$. Hence we have $E|X_1|^2 < \infty$.□

Now we formulate an analogue of the Baum-Katz Theorem (cf. Chow, Teicher (1978), Theorem 5.2.7):

Theorem 3.12 *Let* $|.|$ *a homogeneous norm on* $\mathbb{H}$, *and assume that* $\{X_n\}_{n\geq 1}$ *are i.i.d.* $\mathbb{H}$*-valued random variables. Suppose* $0 < p < 2$, $E|X_1|^p < \infty$, *and let* $c \in \mathbb{H}$ *be such that* $(c', c'') = (-E(X_1)', -E(X_1)'')$ *in case* $1 \leq p < 2$. *Then if* $\alpha p \geq 1$ *we have*

$$\sum_{n=1}^{\infty} n^{\alpha p - 2} P(\max_{1\leq i\leq n} |\delta_{n^{-\alpha}}(\prod_{j=1}^{i}(X_j \cdot c))| \geq \rho) < \infty$$

for every $\rho > 0$.

Proof: The proof is the same as in the classical case (cf. Chow, Teicher (1978), p.130): Again w.l.o.g. $c = 0, E(X_1)' = E(X_1)'' = 0$. By the Marcinkiewicz-Zygmund law of large numbers for $\mathbb{H}$ (cf. Theorem 3.10)

$$\delta_{n^{-1/p}}(\prod_{i=1}^{n} X_i) \overset{a.s.}{\to} 0 \quad (n \to \infty).$$

so

$$\max_{1\leq i\leq n} |\delta_{n^{-1/p}}(\prod_{j=1}^{i} X_j)| \overset{a.s.}{\to} 0 \quad (n \to \infty).$$

Thus

$$\max_{n+1<i<2n} |\delta_{n^{-1/p}}(\prod_{j=n+1}^{i} X_j)|$$

$$= \max_{n+1<i\leq 2n} |\delta_{n^{-1/p}}((-\prod_{j=1}^{n} X_j) \cdot \prod_{j=1}^{i} X_j)|$$

$$\leq C(|\delta_{n^{-1/p}}(\prod_{j=1}^{n} X_j)| + \max_{1<i<2n} |\delta_{(2n)^{-1/p}}(\prod_{j=1}^{i} X_j)|)$$

$$\overset{a.s.}{\to} 0 \quad (n \to \infty). \tag{3.53}$$

Case 1: $\alpha p = 1$. Put

$$\rho' = 2^{-2\alpha}\rho$$

Since the random variables

$$\{\max_{2^n+1\le i<2^{n+1}} |\prod_{j=2^n+1}^{i} X_j|\}_{n\ge 1}$$

are independent, it follows from (3.53) and the Borel-Cantelli Lemma that

$$\begin{aligned}
\infty &> \sum_{n=1}^{\infty} P(\max_{2^n+1\le i\le 2^{n+1}} |\prod_{j=2^n+1}^{i} X_j| \ge 2^{\alpha n}\rho') \\
&= \sum_{n=1}^{\infty} P(\max_{1\le i\le 2^n} |\prod_{j=1}^{i} X_j| \ge 2^{\alpha n}\rho') \\
&\ge \int_0^{\infty} P(\max_{1\le i\le 2^t} |\prod_{j=1}^{i} X_j| \ge 2^{\alpha(t+1)}\rho')dt \\
&\ge (\log 2)^{-1}\int_1^{\infty} x^{-1} P(\max_{1\le i\le x} |\prod_{j=1}^{i} X_j| \ge 2^{\alpha}\rho' x^{\alpha})dx \\
&\ge (\log 2)^{-1}\sum_{n=1}^{\infty} \frac{1}{2n} P(\max_{1\le i\le n} |\prod_{j=1}^{i} X_j| \ge 2^{\alpha}\rho'(2n)^{\alpha}) \\
&= (2\log 2)^{-1}\sum_{n=1}^{\infty} \frac{1}{n} P(\max_{1\le i\le n} |\prod_{j=1}^{i} X_j| \ge \rho n^{\alpha}).
\end{aligned}$$

Case 2: $\alpha p > 1$. Put

$$\rho' = 2^{\alpha^2 p/(\alpha p-1)}\rho.$$

Since for $n \ge 1$

$$\begin{aligned}
(n+1)^{\alpha p/(\alpha p-1)} &\ge n^{\alpha p/(\alpha p-1)} + \frac{\alpha p}{\alpha p-1} n^{1/(\alpha p-1)} \\
&\ge n^{\alpha p/(\alpha p-1)} + n^{1/(\alpha p-1)},
\end{aligned}$$

the random variables

$$\{\max_{n^{\alpha p/(\alpha p-1)}+1\le i\le n^{\alpha p/(\alpha p-1)}+n^{1/(\alpha p-1)}} |\prod_{j=n^{\alpha p/(\alpha p-1)}+1}^{i} X_j|\}_{n\ge 1}$$

are independent, and by (3.53),

$$\begin{aligned}
&\max_{n^{\alpha p/(\alpha p-1)}+1\le i\le n^{\alpha p/(\alpha p-1)}+n^{1/(\alpha p-1)}} n^{-\alpha/(\alpha p-1)}|\prod_{j=n^{\alpha p/(\alpha p-1)}+1}^{i} X_j| \\
&\le \max_{n^{\alpha p/(\alpha p-1)}+1\le i\le 2n^{\alpha p/(\alpha p-1)}} n^{-\alpha/(\alpha p-1)}|\prod_{j=n^{\alpha p/(\alpha p-1)}+1}^{i} X_j| \\
&\overset{a.s.}{\to} 0 \quad (n\to\infty),
\end{aligned}$$

so by the Borel-Cantelli Lemma

$$\begin{aligned}
\infty &> \sum_{n=1}^{\infty} P\Big(\max_{n^{\alpha p/(\alpha p-1)}+1\le i\le n^{\alpha p/(\alpha p-1)}+n^{1/(\alpha p-1)}} \Big|\prod_{j=n^{\alpha p/(\alpha p-1)}+1}^{i} X_j\Big| \\
&\qquad \ge n^{\alpha/(\alpha p-1)}\rho'\Big) \\
&= \sum_{n=1}^{\infty} P\Big(\max_{1\le i\le n^{1/(\alpha p-1)}} \Big|\prod_{j=1}^{i} X_j\Big| \ge n^{\alpha/(\alpha p-1)}\rho'\Big) \\
&\ge \int_1^{\infty} P\Big(\max_{1\le i\le t^{1/(\alpha p-1)}} \Big|\prod_{j=1}^{i} X_j\Big| \ge (t+1)^{\alpha/(\alpha p-1)}\rho'\Big)dt \\
&\ge (\alpha p-1)\int_1^{\infty} x^{\alpha p-2} P\Big(\max_{1\le i\le x} \Big|\prod_{j=1}^{i} X_j\Big| \ge 2^{\alpha/(\alpha p-1)}\rho' x^{\alpha}\Big)dx \\
&\ge A\sum_{n=1}^{\infty} n^{\alpha p-2} P\Big(\max_{1\le i\le n} \Big|\prod_{j=1}^{i} X_j\Big| \ge 2^{\alpha/(\alpha p-1)}\rho'(2n)^{\alpha}\Big) \\
&= A\sum_{n=1}^{\infty} n^{\alpha p-2} P\Big(\max_{1\le i\le n} \Big|\prod_{j=1}^{i} X_j\Big| \ge \rho n^{\alpha}\Big)
\end{aligned}$$

for some constant $A > 0$. □

3.2.3 The ergodic theorem

Let $\{X_n\}_{n\geq 1}$ be a stationary sequence of real-valued random variables on some probability space $(\Omega, \mathcal{F}, P)$ such that $E|X_1| < \infty$. A set $A \in \mathcal{F}$ is called *invariant* if there is a set $B \in \mathcal{B}^{\infty}$ (where $\mathcal{B}$ denotes the Borel σ-algebra on $\mathbb{R}$) such that

$$A = \{\omega \in \Omega : (X_n, X_{n+1}, \dots) \in B\} \quad (n \geq 1).$$

Assume that $\{X_n\}_{n\geq 1}$ is also *ergodic*, i.e. $P(A) \in \{0, 1\}$ for every invariant set A. Now the classical ergodic theorem for stationary sequences says that

$$\frac{1}{n}\sum_{k=1}^{n} X_k \overset{a.s.}{\to} E(X_1) \quad (n \to \infty)$$

(cf. Shiryayev (1984), Theorem V.3.3). The notion of ergodicity carries over to $\mathbb{H}$ in the obvious way.
The following extension is very easy to prove (some further simplifications are owed to an anonymous referee):

Theorem 3.13 *Let $\{X_n\}_{n\geq 1}$ be a stationary ergodic sequence of $\mathbb{H}$-valued random variables on some probability space $(\Omega, \mathcal{A}, P)$ such that $E(X_1)$ exists. Then*

$$\delta_{n^{-1}}\Big(\prod_{k=1}^{n}(X_k \cdot (-E(X_1)))\Big) \overset{a.s.}{\to} 0 \quad (n \to \infty).$$

Proof: Since $E(X_1 \cdot (-E(X_1))) = E(X_1) \cdot (-E(X_1)) = 0$, we may w.l.o.g. assume $E(X_1) = 0$. Since

$$\prod_{k=1}^{n} X_k = \sum_{k=1}^{n} X_k + \frac{1}{2} \sum_{1 \leq k < l \leq n} [X_k, X_l]$$

it suffices, by the classical ergodic theorem for stationary sequences, to prove that

$$n^{-2} \sum_{1 \leq k < l < n} [X_k, X_l] \xrightarrow{a.s.} 0 \quad (n \to \infty).$$

By

$$|n^{-2} \sum_{1 \leq k < l < n} [X_k, X_l]| \leq n^{-1} \sum_{\ell=1}^{n} (|X_\ell| \ell^{-1} |\sum_{k=1}^{\ell-1} X_k|)$$

and an easy limit argument, the assertion follows. □

Remark 3.5 *Clearly, the above proof works also for other laws of large numbers for dependent random variables, as e.g. the law of large numbers for exchangeable sequences (for an elementary proof thereof in the classical case see Pratelli (1989)) and the generalization of the strong law of large numbers by Bose, Chandra (1994) (this is the strong law of large numbers for pairwise independent (or $*$-mixing) mean 0 random variables $\{X_n\}_{n>1}$ such that $\sup_{n>1} \frac{1}{n} \sum_{k=1}^{n} P(|X_k| \geq x)$ is integrable on $[0, \infty[$ and $\sum_{n=1}^{\infty} P(|X_n| \geq n) < \infty$).*

3.2.4 Non-classical laws of the iterated logarithm

Besides the classical Hartman-Wintner law of the iterated logarithm and its (well-known) sharpening by Strassen, there are also several "non-classical" forms of laws of the iterated logarithm available in the literature. See e.g. Bingham (1986) for an excellent survey. One of the most natural versions seems to be the one that has been suggested by Chover (1966): It refers to non-Gaussian stable measures on $\mathbb{R}$. The aim of this section is to show how this can be carried over to $\mathbb{H}$, work which has been done by the author and mainly by Scheffler (cf. Scheffler (1995a, 1995b), Neuenschwander, Scheffler (1996)).

First of all, we prove laws of the iterated logarithm referring to homogeneous norms. Let $|.|$ be the homogeneous norm (1.1) on $\mathbb{H}$. We may assume that the one-parameter automorphism group is given by $\{t^A\}_{t>0} = \{\sigma_{M,m}(t)\}_{t>0}$ $(M \in sp(\mathbb{R}^2), m > 0)$ (this is no loss of generality, since every contracting one-parameter automorphism group is, up to conjugation with an inner automorphism, of this form (cf. 2.1.3)). The following is Theorem 3.1 in Scheffler (1995a).

Theorem 3.14 *Let $X_1, X_2, \ldots,$ be i.i.d. $\mathbb{H}$-valued random variables which obey to an L-S-full probability measure μ which is embeddable into a strictly $\{\sigma_{M,m}\}_{t>0}$-stable c.c.s. without Gaussian component. Then*

$$\limsup_{n \to \infty} |(n \log n)^{-A} (\prod_{j=1}^{n} X_j)|^{(\log \log n)^{-1}} \overset{a.s.}{=} 1.$$

Proof: It suffices to prove that for $\varepsilon > 0$

$$|(n\log n)^{-A}(\prod_{j=1}^{n} X_j)| > (\log n)^{\varepsilon}$$
$$\text{for at the most finitely many } n \text{ a.s.} \tag{3.54}$$

and

$$|(n\log n)^{-A}(\prod_{j=1}^{n} X_j)| > (\log n)^{-\varepsilon}$$
$$\text{for infinitely many } n \text{ a.s.} \tag{3.55}$$

1. Let $n_k := 2^k$,

$$C := (\sup_{k\geq 1} \max_{n_k \leq n \leq n_{k+1}} \|(\frac{n_k \log n_k}{n \log n})^{A}\|)^{-1}$$

(where $\|B\|$ denotes the automorphism norm $\|B\| := \sup_{x\neq 0} \frac{|Bx|}{|x|}$) and define the events

$$A_n := \{|(n\log n)^{-A}(\prod_{j=1}^{n} X_j)| > (\log n)^{\varepsilon}\},$$

$$B_k := \{\max_{n_k \leq n < n_{k+1}} |(n_k \log n_k)^{-A}(\prod_{j=1}^{n} X_j)| > C(\log n_k)^{\varepsilon}\}.$$

With this we get

$$\limsup_{n\to\infty} A_n \subset \limsup_{k\to\infty} B_k. \tag{3.56}$$

Put

$$d' := \max_{n_k \leq n \leq n_{k+1}} P(|(n_k \log n_k)^{-A}(\prod_{j=n+1}^{n_{k+1}} X_j)| > \frac{C}{2}(\log n_k)^{\varepsilon}),$$

$$d := \sup_{0<t<1/\log 2} t^{A}(\mu)(\{x \in \mathbb{H} : |x| > \frac{C}{2}\}).$$

One obtains $d < 1$ and, since $(n_{k+1} - n)/(n_k \log n_k) \in [0, 1/\log 2]$ $(n_k \leq n \leq n_{k+1})$, the strict stability of μ yields $d' \leq d$. By Ottaviani's inequality for homogeneous groups (cf. Scheffler (1995a), Lemma 3.3) we get

$$P(B_k) \leq \frac{1}{1-d} P(|(n_k \log n_k)^{-A}(\prod_{j=1}^{n_{k+1}} X_j)| > \frac{C}{2}(\log n_k)^{\varepsilon}). \tag{3.57}$$

Now pick

$$\beta > \max(\{m\} \cup \{m + \mathrm{Re}\,\lambda : \lambda \in \mathrm{Spec}\, M\}),$$

with which we get

$$\|t^{A}\| \leq \max\{t^{m}, \|t^{M+m}\|\} \leq t^{\beta} \quad (t > 1) \tag{3.58}$$

($\|\cdot\|$ denoting the usual automorphism norm $\|B\| := \sup_{x\neq 0}\frac{\|Bx\|}{\|x\|}$). By the strict stability of μ and the definition of the Lévy measure, we thus obtain, for large k, the estimation

$$
\begin{aligned}
&P(\|(n_k\log n_k)^{-A}(\prod_{j=1}^{n_{k-1}} X_j)\| > \frac{C}{2}(\log n_k)^{\varepsilon}) \\
&= P(\|\delta_{(\log n_k)^{\varepsilon}}((2/\log n_k)^{A}(X_1))\| > \frac{C}{2}) \\
&\leq P(\|(2/(\log n_k)^{1+\varepsilon/\beta})^{A}(X_1)\| > \frac{C}{2}) \\
&\leq Hk^{-1-\varepsilon/\beta}((\log n_k)^{1+\varepsilon/\beta}/2)P(\|(2/(\log n_k)^{1+\varepsilon/\beta})^{A}(X_1)\| > \frac{C}{2}) \\
&\leq Kk^{-1-\varepsilon/\beta}.
\end{aligned}
\tag{3.59}
$$

By (3.56), (3.57), (3.59), and the Borel Cantelli lemma (3.54) follows.
2. In order to prove (3.55), put again $n_k := 2^k$ and define the events

$$E_k := \{\|(n_k\log n_k)^{-A}(\prod_{j=n_{k-1}+1}^{n_k} X_j)\| > (\log n_k)^{-\varepsilon/2}\}.$$

Let β be as above. By (3.58) and the strict stability of μ we get

$$
\begin{aligned}
P(E_k) &= P(\|\delta_{(\log n_k)^{\varepsilon/2}}((2\log n_k)^{-A}(X_1))\| > 1) \\
&\geq P(\|(2(\log n_k)^{1-\varepsilon/(2\beta)})^{-A}(X_1)\| > 1) \\
&\geq Jk^{-1+\varepsilon/(2\beta)}2(\log n_k)^{1-\varepsilon/(2\beta)}P((2(\log n_k)^{1-\varepsilon/(2\beta)})^{-A}(X_1) > 1).
\end{aligned}
$$

So by the definition of the Lévy measure

$$P(E_k) \geq Lk^{-1+\varepsilon/(2\beta)} \quad (k \geq 0). \tag{3.60}$$

Now, since the E_k are independent, the Borel-Cantelli lemma yields

$$P(\limsup_{k\to\infty} E_k) = 1. \tag{3.61}$$

Now suppose (3.55) is false with positive probability. Then by (3.61)

$$
\begin{aligned}
(\log n_{k+1})^{-\varepsilon} &\geq \|(n_{k+1}\log n_{k+1})^{-A}(\prod_{j=1}^{n_k} X_j)\cdot(n_{k+1}\log n_{k+1})^{-A}(\prod_{j=n_k+1}^{n_{k+1}} X_j)\| \\
&> (\log n_{k+1})^{-\varepsilon/2} - \|(\frac{n_k\log n_k}{n_{k+1}\log n_{k+1}})^{A}\|(\log n_k)^{-\varepsilon}
\end{aligned}
$$

for infinitely many k with positive probability. (3.62)

But, since for large k the right hand side of (3.62) is bigger than $(\log n_{k+1})^{-\varepsilon}$, we get a contradiction. □

One can prove the following (probabily still incomplete) complement to Theorem 3.14 (cf. Scheffler (1995a), Corollary 3.4):

Theorem 3.15 *Let $\lambda \in]0,1[$. Then in the situation of Theorem 3.14, 1 is a.s. an accumulation point of*

$$\{|(n(\log n)^{\lambda})^{-A}(\prod_{j=1}^{n} X_j)|^{(\log\log n)^{-1}}\}_{n>1}.$$

Proof: Take the sequence $n_k = \lfloor 2^{k^{1/\lambda}} \rfloor$ $(k \geq 1)$. We have to show

$$|n_k^{-A}(\prod_{j=1}^{n_k} X_j)| > (\log n_k)^{(1+\varepsilon)\lambda} \text{ for at most finitely many } k \text{ a.s.} \tag{3.63}$$

and

$$|n_k^{-A}(\prod_{j=1}^{n_k} X_j)| > (\log n_k)^{(1-\varepsilon)\lambda} \text{ for infinitely many } k \text{ a.s.} \tag{3.64}$$

By the strict stability of X_1 and the definition of the Lévy measure, we get (similarly as in (3.59))

$$P(|n_k^{-A}(\prod_{j=1}^{n_k} X_j| > (\log n_k)^{(1+\varepsilon)\lambda}) \leq K k^{-(1+\rho)}$$

for some $\rho, K > 0$. Now the Borel-Cantelli lemma yields (3.63). In order to prove (3.64), consider the events

$$E_k := \{|n_k{}^{A}(\prod_{j=n_{k-1}+1}^{n_k} X_j)| > (\log n_k)^{(1-\varepsilon/2)\lambda}\}.$$

Again by the strict stability and the definition of the Lévy measure we obtain (similarly as in (3.60))

$$\begin{aligned} P(E_k) &= P(|(\frac{n_k}{n_k - n_{k-1}})^{-A}(X_1)| > (\log n_k)^{(1-\varepsilon/2)\lambda}) \\ &\geq K k^{-1+\rho'} \end{aligned}$$

for some $\rho' > 0$. Hence by the Borel-Cantelli lemma

$$P(\limsup_{k\to\infty} E_k) = 1. \tag{3.65}$$

Now suppose (3.64) is false with positive probability. Then the contradiction follows similarly as in (3.62). □

Related theorems for the semistable case are the following (cf. Scheffler (1995b), Theorem 3.4).

For the definition of $\mathcal{B} \subset Aut(\mathbb{H})$ cf. 2.1.3.

Theorem 3.16 *Let $\tau \in \mathcal{B}$, $\{\mu_t\}_{t>0}$ a strictly (τ, c)-semistable c.c.s. on $\mathbb{H}$ with Lévy measure $\eta \neq 0$, and $X_1, X_2, \ldots$ i.i.d. $\mathbb{H}$-valued random variables obeyiang to the law μ_1. If the μ_t are L-S-full or if $q(\eta) \neq 0$, then*

$$\limsup_{n\to\infty} |(\tau^{-(n-\lfloor\frac{\log n}{\log c}\rfloor)}(\prod_{j=1}^{c^n} X_j)|^{(\log n)^{-1}} \overset{a.s.}{=} 1. \tag{3.66}$$

Also here, we have an analogue of Theorem 3.15:

Theorem 3.17 *In the situation of Theorem 3.16, for* $\lambda \in]0,1[$*, the value* 1 *is a.s. an accumulation point of*

$$\{|r^{-(n+\lambda \frac{\log n}{\log c})}(\prod_{j=1}^{c^n} X_j)|^{(\log n)^{-1}}\}_{n>1}.$$

The proofs of Theorems 3.16 and 3.17, being similar to the stable case and to the corresponding statements for the euclidean case, are omitted.
Now we are going to formulate analogous statements for the central part $q(\prod_{j=1}^n X_j)$ (cf. Neuenschwander, Scheffler (1996)).
First, we establish a lemma for probability measures on $\mathbb{R}$ lying in the strict domain of normal attraction of a strictly $\{t^{1/\alpha}\}_{t>0}$-stable measure ($\alpha \in]0,1[$) (which is also of independent interest).

Lemma 3.8 *Let* $\{t^{1/\alpha}\mu\}_{t\geq 0}$ *be a strictly* $\{t^{1/\alpha}\}_{t>0}$*-stable semigroup on* $\mathbb{R}$ *with* $\alpha \in]0,1[$*. Furthermore let* $\nu \in M^1(\mathbb{R})$ *be symmetric and assume* $\nu \in SDONA(\{t^{1/\alpha}\mu\}_{t>0},]0,\infty[)$*. Let* $X_1, X_2, \ldots$ *be an i.i.d. sequence of real valued random variables obeying to the law* ν*. For* $0 < \lambda \leq 1$ *let*

$$n_k := \lfloor 2^{k^{1/\lambda}} \rfloor.$$

Then we have

$$\limsup_{n\to\infty} |n_k^{-1/\alpha}(\log n_k)^{-\lambda/\alpha} \sum_{j=1}^{n_k} X_j|^{(\log\log n_k)^{-1}} \overset{a.s.}{=} 1.$$

Proof: It suffices to show that for fixed $\varepsilon > 0$, we have

$$|\sum_{j=1}^{n_k} X_j| > n_k^{1/\alpha}(\log n_k)^{\lambda/\alpha+\varepsilon} \quad \text{for finitely many } k \text{ a.s.} \tag{3.67}$$

and

$$|\sum_{j=1}^{n_k} X_j| > n_k^{1/\alpha}(\log n_k)^{\lambda/\alpha-\varepsilon} \quad \text{for infinitely many } k \text{ a.s.} \tag{3.68}$$

1. By the well-known convergence criteria for triangular systems of probability measures on $\mathbb{R}$ we get

$$n(n^{-1/\alpha}(\nu))|_{cpl\,U} \xrightarrow{w} \eta|_{cpl\,U}$$

for any Borel neighborhood U of 0 which is a continuity set of η, where η is the Lévy measure of $\{t^{1/\alpha}\mu\}_{t\geq 0}$. Then from the strict stability property $t^{1/\alpha}(\eta) = t\cdot\eta$ we get

$$P(|X_1| > t\} \sim t^{-\alpha} \quad (t \to \infty). \tag{3.69}$$

In the following let

$$x(n) := n^{1/\alpha}(\log n)^{\lambda/\alpha+\varepsilon}.$$

By Araujo, Giné (1980), Theorem 2.6.17 we see that condition (1) of Heyde (1967) is fulfilled, whereas his condition (2) is immediate from (3.69). Furthermore it is easy to check, using the Lemma in Heyde (1967), that

$$x(n)^{-1}\sum_{j=1}^{n} X_j \xrightarrow{w} 0.$$

So in view of the Theorem in Heyde (1967) there exists a constant $C > 0$ such that for n large enough we have

$$P(|\sum_{j=1}^{n} X_j| > x(n)) \leq CnP(|X_1| > x(n)). \tag{3.70}$$

By (3.69) and (3.70) we get

$$\begin{aligned} P(|\sum_{j=1}^{n_k} X_j| > n_k^{1/\alpha}(\log n_k)^{\lambda/\alpha+\varepsilon}) &\leq Cn_kP(|X_1| > n_k^{1/\alpha}(\log n_k)^{\lambda/\alpha+\varepsilon}) \\ &\leq Ck^{-1-\varepsilon\alpha/\lambda}. \end{aligned}$$

Now an application of the Borel-Cantelli lemma yields (3.67).
2. In order to prove (3.68) for small $\varepsilon > 0$ put

$$y(n) := (2n)^{1/\alpha}(\frac{\log(2n)}{\log 2})^{\lambda/\alpha-\varepsilon/2}.$$

Another application of the Theorem in Heyde (1967) this time gives for large n and some constant $D > 0$

$$P(|\sum_{j=1}^{n} X_j| > y(n)) \geq DnP(|X_1| > y(n)). \tag{3.71}$$

First we show that we have

$$|\sum_{j=n_{k-1}+1}^{n_k} X_j| > n_k^{1/\alpha}(\log n_k)^{\lambda/\alpha-\varepsilon/2} \quad \text{for infinitely many } k \text{ a.s.} \tag{3.72}$$

Note that since $0 < \lambda \leq 1$ there exist $C_1, C_2 > 0$ such that $C_1 n_k \leq n_k - n_{k-1} \leq C_2 n_k$ for all large k. Using (3.69) and (3.71), we have for large k and some $C, D, D' > 0$ that

$$\begin{aligned} P(|\sum_{j=n_{k-1}+1}^{n_k} X_j| > n_k^{1/\alpha}(\log n_k)^{\lambda/\alpha-\varepsilon/2}) &= P(|\sum_{j=1}^{n_k-n_{k-1}} X_j| > n_k^{1/\alpha}(\log n_k)^{\lambda/\alpha-\varepsilon/2}) \\ &\geq P(|\sum_{j=1}^{n_k-n_{k-1}} X_j| > Cy(n_k - n_{k-1})) \\ &\geq D(n_k - n_{k-1})P(|X_1| > Cy(n_k - n_{k-1})) \\ &\geq DC_1n_kP(|X_1| > Cy(C_2n_k)) \\ &\geq D'k^{-1+\varepsilon\alpha/(2\lambda)}. \end{aligned}$$

The independence part of the Borel-Cantelli lemma now yields (3.72). Now suppose that (3.68) is false on a set of positive probability. Then, using (3.72) and the already proved upper bound (3.67), we get

$$\begin{aligned} n_k^{1/\alpha}(\log n_k)^{\lambda/\alpha+\varepsilon} &> |\sum_{j=1}^{n_k} X_j| \\ &\geq |\sum_{j=n_{k-1}+1}^{n_k} X_j| - |\sum_{j=1}^{n_{k-1}} X_j| \\ &> n_k^{1/\alpha}(\log n_k)^{\lambda/\alpha-\varepsilon/2} - n_{k-1}^{1/\alpha}(\log n_{k-1})^{\lambda/\alpha+\varepsilon} \end{aligned}$$

for infinitely many k with positive probability.

But for large k the last difference is bigger than $n_k^{1/\alpha}(\log n_k)^{\lambda/\alpha+\varepsilon}$, which is a contradiction. □

Theorem 3.18 *Assume $X_1, X_2, \ldots$ are i.i.d. L-S-full symmetric $\mathbb{H}$-valued random variables whose law is embeddable into a strictly $\{t^A\}_{t>0}$-stable c.c.s. without Gaussian component and with $\{t^A\}_{t>0} = \{\sigma_{M,m}(t)\}_{t>0}$ ($M \in sp(\mathbb{R}^2), m \geq \frac{1}{2}$) (cf. 2.1.3). Suppose the Lévy measure η of $\{t^A \mathcal{L}(X_1)\}_{t>0}$ satisfies $q(\eta) \not\equiv 0$. Then*

$$\limsup_{n\to\infty} |q((n\log n)^{-A}(\prod_{j=1}^{n} X_j))|^{(\log\log n)^{-1}} \overset{a.s.}{=} 1.$$

Proof: 1. The "$\leq$"-direction follows at once from the corresponding result for $|\prod_{j=1}^n X_j|$ (Theorem 3.14).
2. Now we prove the "$\geq$"-part. Assume the contrary, which by the Kolmogorov 0-1 law reads as

$$\limsup_{n\to\infty} |q((n\log n)^{-A}(\prod_{j=1}^{n} X_j))|^{(\log\log n)^{-1}} \overset{a.s.}{=} e^{-2\delta}$$

for some (fixed) $\delta > 0$. Hence

$$|q(\prod_{j=1}^{n} X_j)| \leq n^{2m}(\log n)^{2m-\delta} \quad \text{for all but a finite number of } n \text{ a.s.} \tag{3.73}$$

By the symmetry, both processes

$$\{\prod_{j=1}^{n} X_j\}_{n\geq 1}, \{-\prod_{j=1}^{n} X_{n+1-j}\}_{n\geq 1}$$

have the same distribution, so the same holds for the pair of processes

$$\{|q(\prod_{j=1}^{n} X_j)|\}_{n\geq 1}, \{|q(\prod_{j=1}^{n} X_{n+1-j})|\}_{n\geq 1},$$

thus (3.73) implies also

$$|q(\prod_{j=1}^{n} X_{n+1-j})| \leq n^{2m}(\log n)^{2m-\delta} \quad \text{for all but a finite number of } n \text{ a.s.} \tag{3.74}$$

By (3.73) and (3.74)

$$|q(\frac{1}{2}(\prod_{j=1}^{n} X_j + \prod_{j=1}^{n} X_{n+1-j}))| \leq n^{2m}(\log n)^{2m-\delta} \quad \text{for all but a finite number of } n \text{ a.s.} \tag{3.75}$$

But as in (2.16)

$$\prod_{j=1}^{n} X_j + \prod_{j=1}^{n} X_{n+1-j} = 2\sum_{j=1}^{n} X_j,$$

so it follows from (3.75) that

$$|q(\sum_{j=1}^{n} X_j)| \leq n^{2m}(\log n)^{2m-\delta} \quad \text{for all but a finite number of } n \text{ a.s.}$$

Hence

$$\limsup_{n\to\infty} |(n \log n)^{-2m} \sum_{j=1}^{n} q(X_j)|^{\frac{1}{\log\log n}} \overset{a.s.}{\leq} e^{-\delta} < 1. \tag{3.76}$$

A slight variant of Lemma 2.4 (here (2.12) holds trivially), together with the fact that $q(\eta) \not\equiv 0$, yields that the symmetric law $q(X_1)$ lies in the strict domain of normal attraction of a non-normal non-degenerate strictly $\{t^{2m}\}$-stable c.c.s. on $I\!R$. But now (since $m \geq \frac{1}{2}$) (3.76) contradicts Lemma 3.8 (with $\lambda = 1$). □

Now we are going to formulate a complement to Theorem 3.18 analogous to Theorem 3.15:

Theorem 3.19 *Let $\lambda \in]0,1[$. Then in the situation of Theorem 3.18 we have that 1 is a.s. an accumulation point of*

$$\{|q((n(\log n)^{\lambda})^{-A}(\prod_{j=1}^{n} X_j))|^{(\log\log n)^{-1}}\}_{n\geq 1}.$$

Proof: Put

$$n_k := \lfloor 2^{k^{1/\lambda}} \rfloor.$$

We prove

$$\limsup_{k\to\infty} |q((n_k(\log n_k)^{\lambda})^{-A}(\prod_{j=1}^{n_k} X_j))|^{(\log\log n_k)^{-1}} \overset{a.s.}{=} 1.$$

1. The " $\leq$ "-direction follows again immediately from the proof of Theorem 3.15.
2. As far as the " $\geq$ "-assertion is concerned, the proof is similar to that of Theorem 3.18: Assume the contrary, which again by the Kolmogorov 0-1 law reads as

$$\limsup_{k\to\infty} |q((n_k(\log n_k)^{\lambda})^{-A}(\prod_{j=1}^{n_k} X_j))|^{(\log\log n_k)^{-1}} \overset{a.s.}{=} e^{-2\delta}$$

for some $\delta > 0$. Hence

$$|q((n_k(\log n_k)^{\lambda})^{-A}(\prod_{j=1}^{n_k} X_j))|^{(\log\log n_k)^{-1}} \leq e^{-\delta} \quad \text{for all but a finite number of } k \text{ a.s.}$$

So

$$|q(\prod_{j=1}^{n_k} X_j)| \le n_k^{2m}(\log n_k)^{2\lambda m-\delta} \quad \text{for all but a finite number of } k \text{ a.s.}$$

As in the proof of Theorem 3.18 it follows that

$$|q(\sum_{j=1}^{n_k} X_j)| \le n_k^{2m}(\log n_k)^{2\lambda m-\delta} \quad \text{for all but a finite number of } k \text{ a.s.,}$$

which contradicts Lemma 3.8. □

Now assume $\tau \in Aut(\mathbb{H})$ is of the form

$$\tau = \begin{pmatrix} mM & 0 \\ 0 & [-]m^2 \end{pmatrix}$$

($m > 0$, M [skew-]symplectic). By the proof of Scheffler (1995b), Lemma 3.3 we have $m > 1$ if there exists a strictly (τ, c)-semistable c.c.s $\{\mu_t\}_{t>0}$ whose Lévy measure η satisfies $q(\eta) \not\equiv 0$.

Theorem 3.20 *Let $\{\mu_t\}_{t>0}$ be an L-S-full strictly (τ, c)-semistable c.c.s. without Gaussian component and with Lévy measure η, τ as above, μ_t symmetric $(t \ge 0)$, $q(\eta) \not\equiv 0$. Assume $X_1, X_2, \ldots$ are i.i.d. $\mathbb{H}$-valued random variables with $\mathcal{L}(X_1) = \mu_1$. Then*

$$\limsup_{n\to\infty} |q(\tau^{-(n+\lfloor\frac{\log n}{\log c}\rfloor)}(\prod_{j=1}^{c^n} X_j))|^{(\log n)^{-1}} \overset{a.s.}{=} 1.$$

Proof: 1. The " $\le$ "-part follows at once from Theorem 3.16.
2. For the " $\ge$ "-part we again adapt the method of proof of Theorem 3.18: Assume the contrary, which by the Kolmogorov 0-1 law reads as

$$\limsup_{n\to\infty} |q(\tau^{-(n+\lfloor\frac{\log n}{\log c}\rfloor)}(\prod_{j=1}^{c^n} X_j))|^{(\log n)^{-1}} \overset{a.s.}{=} e^{-2\delta}$$

for some $\delta > 0$, hence

$$|q(\tau^{-(n+\lfloor\frac{\log n}{\log c}\rfloor)}(\prod_{j=1}^{c^n} X_j))|^{(\log n)^{-1}} \le e^{-\delta} \quad \text{for all but a finite number of } n \text{ a.s.,}$$

which implies

$$|q(\prod_{j=1}^{c^n} X_j)| \le n^{-\delta} m^{2(n+\lfloor\frac{\log n}{\log c}\rfloor)} \quad \text{for all but a finite number of } n \text{ a.s.}$$

As in the proof of Theorem 3.18 it follows that

$$\limsup_{n\to\infty} |q(m^{-2(n+\lfloor\frac{\log n}{\log c}\rfloor)}(\sum_{j=1}^{c^n} X_j))|^{(\log n)^{-1}} \overset{a.s.}{\le} e^{-\delta}, \tag{3.77}$$

but since $q(X_1)$ is symmetric and lies in the strict domain of normal attraction of a strictly (m^2, c)-semistable measure on $\mathbb{R}$ with Lévy measure $q(\eta) \not\equiv 0$ and without Gaussian component (by the same variant of Lemma 2.4 as in the proof of Theorem 3.18), this contradicts Scheffler (1995b), Theorem 2.5.□

Also here, we can prove an analogue of Theorem 3.15:

Theorem 3.21 *Let $\lambda \in]0,1[$. Then, in the situation of Theorem 3.20, the value 1 is a.s. an accumulation point of*

$$\{|q(\tau^{-(n+\lfloor\lambda\frac{\log n}{\log c}\rfloor)}(\sum_{j=1}^{c^n} X_j))|^{(\log n)^{-1}}\}_{n\geq 1}.$$

Proof: Put $n_k := \lfloor k^{1/\lambda}\rfloor$. Then one shows

$$\limsup_{k\to\infty} |q(\tau^{-(n_k+\lambda\frac{\log n_k}{\log c})}(\sum_{j=1}^{c^{n_k}} X_j))|^{(\log n_k)^{-1}} \overset{a.s.}{=} 1$$

as in the proof of Theorem 3.20. □

Remark 3.6 *Whereas Theorems 3.14, 3.15, 3.18, and 3.19 are limit theorems for the whole sequence $\{n\}_{n\geq 1}$, Theorems 3.16, 3.17, 3.20, and 3.21 refer to subsequences $\{\lfloor c^n\rfloor\}_{n\geq 1}$. Therefore, though (strictly) stable measures are special cases of (strictly) semistable ones, Theorems 3.16, 3.17, 3.20, and 3.21 yield also new limit theorems for stable measures.*

Similar theorems are also obtainable for general positively graduated simply connected nilpotent Lie groups. However, here the admissible automorphisms are, up to now, only the dilatations δ_t. The carrying over tor more general automorphisms remains open. Cf. Scheffler (1995a), (1995b).

3.2.5 The two-series theorem

The aim of this section is to carry over Kolmogorov's classical Three-Series theorem (see e.g. Shiryayev (1984), Theorem IV.2.3) to symmetric random variables on $\mathbb{H}$. For a $\mathbb{H}$-valued random variable X and $c > 0$, let $X_c := X \cdot \mathbf{1}_{\{\|X\|<c\}}$.

Theorem 3.22 *Let $\{X_n\}_{n\geq 1}$ be a sequence of independent symmetric $\mathbb{H}$-valued random variables. If*

$$\sum_{n=1}^{\infty} E\|X_{n,c}\|^2 < \infty \tag{3.78}$$

and

$$\sum_{n=1}^{\infty} P(\|X_n\| \geq c) < \infty \tag{3.79}$$

for some $c > 0$, then $\coprod_{n=1}^{\infty} X_n$ is a.s. convergent. Conversely, if $\coprod_{n=1}^{\infty} X_n$ is a.s. convergent, then (3.78) and (3.79) hold for every $c > 0$.

Remark 3.7 *Observe that by the fact that $\mathbb{H}$ has no non-trivial compact subgroups (a fact which was already mentioned in 1.1), the a.s. convergence of $\coprod_{n=1}^{\infty} X_n$ is equivalent to the weak (and thus also to the stochastic) convergence (cf. Heyer (1977), Theorem 2.2.19).*

Proof of Theorem 3.22: 1. (Cf. Neuenschwander, Scheffler (1995).) Assume $\prod_{n=1}^{\infty} X_n$ is a.s. convergent. As in the proof of Theorem 3.18, by the symmetry the processes

$$\{\prod_{n=1}^{N} X_n\}_{N\geq 1} \quad , \quad \{-\prod_{n=1}^{N} X_{N+1-n}\}_{N\geq 1}$$

have the same distribution, so

$$\{-\prod_{n=1}^{N} X_{N+1-n}\}_{N\geq 1}$$

is a Cauchy sequence, hence

$$\lim_{N\to\infty}(\prod_{n=1}^{N} X_{N+1-n})$$

and thus

$$\lim_{N\to\infty}(\prod_{n=1}^{N} X_n + \prod_{n=1}^{N} X_{N+1-n})$$

exists a.s. But as in (2.16)

$$\prod_{n=1}^{N} X_n + \prod_{n=1}^{N} X_{N+1-n} = 2\sum_{n=1}^{N} X_n,$$

hence (3.78) and (3.79) hold for every $c > 0$ by the classical Three-Series theorem for $(\mathbb{R}^3, +)$.

2. The other direction has been proved by Neuenschwander, Scheffler (1995) for all simply connected nilpotent Lie groups by Fourier analytic methods. Here we give a different proof based on martingale theory. Suppose (3.78) and (3.79) hold for some $c > 0$. Since by (3.79) and the Borel-Cantelli Lemma

$$P(X_n \neq X_{n,c} \textit{ for infinitely many } n) = 0,$$

it suffices to prove that $\prod_{n=1}^{\infty} X_{n,c}$ is a.s. convergent. By the symmetry

$$E(X_{n,c}) = 0, \tag{3.80}$$

so we have a.s.

$$\begin{aligned} E(\prod_{n=1}^{N} X_{n,c} \mid \prod_{n=1}^{N-1} X_{n,c}) &= E(\prod_{n=1}^{N-1} X_{n,c} + X_{N,c} + \frac{1}{2}[\prod_{n=1}^{N-1} X_{n,c}, X_{N,c}] \mid \prod_{n=1}^{N-1} X_{n,c}) \\ &= \prod_{n=1}^{N-1} X_{n,c}, \end{aligned}$$

so $\{\prod_{n=1}^{N} X_{n,c}\}_{N\geq 1}$ is a martingale. Hence by Doob's convergence theorem (cf. Shiryayev (1984), Theorem VII.4.1) it suffices to show that

$$\sup_{N\geq 1} E\|\prod_{n=1}^{N} X_{n,c}\| < \infty. \tag{3.81}$$

We prove the stronger assertion

$$\sup_{N\geq 1} E\,\|\prod_{n=1}^{N} X_{n,c}\|^2 < \infty. \tag{3.82}$$

Indeed, by (3.78) and (3.80) there is a $K \geq 0$ such that

$$\begin{aligned}
&\sup_{N\geq 1} E\|\prod_{n=1}^{N} X_{n,c}\|^2 \\
&= \sup_{N\geq 1} E\,\|\sum_{n=1}^{N} X_{n,c} + \frac{1}{2}\sum_{1\leq n<m\leq N}[X_{n,c}, X_{m,c}]\|^2 \\
&\leq \sup_{N\geq 1}\{\sum_{n=1}^{N} E\|X_{n,c}\|^2 + K\sum_{1\leq n<m\leq N} E\|X_{n,c}\|^2 E\|X_{m,c}\|^2 \\
&\quad +2\sum_{1\leq n<n'\leq N} E\langle X_{n,c}, X_{n',c}\rangle + \sum_{n=1}^{N}\sum_{1\leq n'<m'\leq N} E\langle X_{n,c}, [X_{n',c}, X_{m',c}]\rangle \\
&\quad +\frac{1}{2}\sum_{1\leq n<n'\leq N,\, n<m\leq N,\, n'<m'\leq N} E\langle [X_{n,c}, X_{m,c}], [X_{n',c}, X_{m',c}]\rangle \\
&\quad +\frac{1}{2}\sum_{n=1}^{N}\sum_{n<m<m'} E\langle [X_{n,c}, X_{m,c}], [X_{n,c}, X_{m',c}]\rangle\} \\
&= \sup_{N\geq 1}\{\sum_{n=1}^{N} E\|X_{n,c}\|^2 + K\sum_{1\leq n<m\leq N} E\|X_{n,c}\|^2 E\|X_{m,c}\|^2\} \\
&\leq \sup_{N\geq 1}\{\sum_{n=1}^{N} E\|X_{n,c}\|^2 + \frac{K}{2}(\sum_{n=1}^{N} E\|X_{n,c}\|^2)^2\} \\
&< \infty
\end{aligned}$$

(observe that in any of the expressions $E\langle\ldots\rangle$ at least one of the occurring indexes is different from any of the others, so that by the independence of the $X_{n,c}$ and (3.80) the expression gets 0). So (3.82) and thus (3.81) is proved. □

Bibliography

[1] Aldous, D. J. (1977). Limit theorems for subsequences of arbitrarily-dependent sequences of random variables. *Z. Wahrscheinlichkeitstheorie verw. Geb.* **40**, 59-82.

[2] Araujo, A., Giné, E. (1980). *The central limit theorem for real and Banach valued random variables*. Wiley, New York.

[3] Azencott, R. (1970). *Espaces de poisson des groupes localement compacts*. Lecture Notes in Mathematics 148. Springer, Berlin.

[4] Baldi, P. (1985). Unicité du plongement d'une mesure de probabilité dans un semi-groupe de convolution gaussien. Cas non-abélien. *Math. Z.* **188**, 411-417.

[5] Baldi, P. (1986). Large deviations and functional iterated logarithm law for diffusion processes. *Probab. Th. Rel. Fields* **71**, 435-453.

[6] Baldi, P. (1990). Large deviations and the functional Lévy's modulus for invariant diffusions. In: Korezlioglu, H., Ustunel, A. S. (ed.). *Stochastic analysis and related topics II*. Lecture Notes in Mathematics 1444. Springer, Berlin, 193-203.

[7] Béguin, C., Valette, A., Zuk, A. (1995). *On the spectrum of a random walk on the discrete Heisenberg group and the norm of Harper's operator*. Preprint.

[8] Berthuet, R. (1979). Loi du logarithme itéré pour certaines intégrales stochastiques. *C. R. Acad. Sci. Paris A* **289**, 813-815.

[9] Berthuet, R. (1986). Etude de processus généralisant l'aire de Lévy. *Probab. Th. Rel. Fields* **73**, 464-480.

[10] Billingsley, P. (1968). *Convergence of probability measures*. Wiley, New York.

[11] Bingham, M. S. (1993). Convergence of approximate martingale arrays to mixtures of infinitely divisible distributions in locally compact abelian groups. *J. Theoret. Probab.* **6(1)**, 123-152.

[12] Bingham, N. H. (1986). Variants on the law of the iterated logarithm. *Bull. London Math. Soc.* **18**, 433-467.

[13] Bismut, J. M. (1984). The Atiyah-Singer theorems. *J. Funct. Anal.* **57**, 56-99 and 329-348.

[14] Bismut, J. M. (1988). Formules de localisation et formules de Paul Lévy. *Astérisque* **157-158**, 37-58.

[15] Bloom, W. R., Heyer, H. (1995). *Harmonic analysis of probability measures on hypergroups*. De Gruyter, Berlin.

[16] Blumenthal, R. M., Getoor, R. K. (1968). *Markov processes and potential theory*. Academic Press, New York.

[17] Böge, W. (1959). Ueber die Charakterisierung unendlich teilbarer Wahrscheinlichkeitsverteilungen. *J. Reine Angew. Math.* **201**, 150-156.

[18] Böge, W. (1964). Zur Charakterisierung sukzessiv unendlich teilbarer Wahrscheinlichkeitsverteilungen. *Z. Wahrscheinlichkeitstheorie verw. Geb.* **2**, 380-394.

[19] Bony, J. M. (1969). Principe du maximum, inégalité de Harnack et unicité du problème de Cauchy pour les opérateurs elliptiques dégénérés. *Ann. Inst. Fourier (Grenoble)* **19(1)**, 277-304.

[20] Bose, A., Chandra, T. K. (1994). A note on the strong law of large numbers. *Calcutta Statist. Assoc. Bull.* **44(173-174)**, 115-122.

[21] Breiman, L. (1968). *Probability*. Addison-Wesley, Reading (Mass.).

[22] Burrell, Q. L., McCrudden, M. (1974). Infinitely divisible distributions on connected nilpotent Lie groups. *J. London Math. Soc. (2)* **7**, 584-588.

[23] Carnal, H. (1966). Deux théorèmes sur les groupes stochastiques compacts. *Comm. Math. Helv.* **40**, 237-246.

[24] Carnal, H. (1986). Les variables aléatoires de loi stable et leur représentation selon P. Lévy. In: Heyer, H. (ed.). *Probability measures on groups VIII*. Lecture Notes in Mathematics 1210. Springer, Berlin, 24-33.

[25] Chaleyat-Maurel, M. (1981). Densités des diffusions invariantes sur certains groupes nilpotents. Calcul d'après B. Gaveau. *Astérisque* **84-85**, 203-214.

[26] Chaleyat-Maurel, M., Le Gall, J.-F. (1989). Green function, capacity, and sample path properties for a class of hypoelliptic diffusion processes. *Probab. Theory Related Fields* **83**, 219-264.

[27] Chatterji, S. D. (1969). An L^p-convergence theorem. *Ann. Math. Stat.* **40(3)**, 1068-1070.

[28] Chatterji, S. D. (1974). A subsequence principle in probability theory. *Bull. Amer. Math. Soc.* **80**, 495-497.

[29] Chover, J. (1966). A law of the iterated logarithm for stable summands. *Proc. Amer. Math. Soc.* **17**, 441-443.

[30] Chow, Y. S., Teicher, H. (1978). *Probability theory*. Springer, Berlin.

[31] Chung, K. L. (1948). On the maximum partial sum of sequences of independent random variables. *Trans. Amer. Math. Soc.* **64**, 205-233.

[32] Coşkun, E. *Faltungshalbgruppen von Wahrscheinlichkeitsmassen auf einer Hilbert-Lie-Gruppe*. Dissertation, University of Tübingen.

[33] Crépel, P., Raugi, A. (1978). Théorème central limite sur les groupes nilpotents. *Ann. Inst. H. Poincaré Probab. Statist.* **14**, 145-162.

[34] Crépel, P., Roynette, B. (1977). Une loi du logarithme itéré pour le groupe d'Heisenberg. *Z. Wahrscheinlichkeitstheorie verw. Geb.* **39**, 217-229.

[35] Csörgö, M., Révész, P. (1981). *Strong approximations in probability and statistics*. Academic Press, New York.

[36] Dani, S. G. (1991). A characterization of Cauchy-type distributions on boundaries of semisimple groups. *J. Theoret. Probab.* **4**, 625-629.

[37] Doeblin, W. (1940). Sur l'ensemble de puissances d'une loi de probabilité. *Studia Math.* **9**, 71-96.

[38] Drisch, T., Gallardo, L. (1984). Stable laws on the Heisenberg groups. In: Heyer, H. (ed.). *Probability measures on groups VII*. Lecture Notes in Mathematics 1064. Springer, Berlin, 56-79.

[39] Dunau, J. L., Sénateur, H. (1986). Une caractérisation du type de la loi de Cauchy-Heisenberg. In: Heyer, H. (ed.). *Probability measures on groups VIII*. Lecture Notes in Mathematics 1210. Springer, Berlin, 41-57.

[40] Dunau, J. L., Sénateur, H. (1988). A characterization of the type of the Cauchy-Hua measure on real symmetric matrices. *J. Theoret. Probab.* **1(3)**, 263-270.

[41] Feinsilver, Ph. (1978). Processes with independent increments on a Lie group. *Trans. Amer. Math. Soc.* **242**, 73-121.

[42] Feinsilver, Ph., Schott, R. (1989). An operator approach to processes on Lie groups. In: Cambanis, S., Weron, A. (ed.). *Probability theory on vector spaces IV*. Lecture Notes in Mathematics 1391. Springer, Berlin, 59-65.

[43] Feinsilver, Ph., Schott, R. (1996). *Algebraic structures and operator calculus. Vol. III: Representations of Lie groups*. Kluwer, Dordrecht.

[44] Feinsilver, Ph., Franz, U., Schott, R. (1995a). Feynman-Kac formula and Appell systems on quantum groups. *C. R. Acad. Sci. Paris I* **321**, 1615-1619.

[45] Feinsilver, Ph., Franz, U., Schott, R. (1995b). *Duality and multiplicative stochastic processes on quantum groups*. Prépublication de l'Institut Elie Cartan **95/26**, Nancy.

[46] Fel'dman, G. (1987). Bernstein Gaussian distributions on groups. *Theory Probab. Appl.* **31(1)**, 40-49.

[47] Feller, W. (1968). *An introduction to probability theory and its applications.* Vol. I, 3rd ed. Wiley, New York.

[48] Folland, G. B. (1989). *Harmonic analysis in phase space.* Princeton University Press, Princeton NJ.

[49] Folland, G. B., Stein, E. M. (1982). *Hardy spaces on homogeneous groups.* Princeton University Press, Princeton NJ.

[50] Franz, U., Schott, R. (1996). *Diffusions on braided spaces.* Prépublication de l'Institut Elie Cartan **96/4**, Nancy.

[51] Freidlin, M. I., Ventsel, A. D. (1984). *Random perturbations of dynamical systems.* Springer, Berlin.

[52] Furstenberg, H. (1963). Noncommuting random products. *Trans. Amer. Math. Soc.* **108**, 377-428.

[53] Gallardo, L. (1982). Capacités, mouvement Brownien et problème de l'épine de Lebesgue sur les groupes de Lie nilpotents. In: Heyer, H. (ed.). *Probability measures on groups.* Lecture Notes in Mathematics 928. Springer, Berlin, 96-120.

[54] Gangolli, R. (1964). Isotropic infinitely divisible measures on symmetric spaces. *Acta Math.* **111**, 213-246.

[55] Gapoškin, V. F. (1972). Convergence and limit theorems for sequences of random variables. *Th. Probab. Appl.* **17**, 379-399.

[56] Gaveau, B. (1977). Principe de moindre action, propagation de la chaleur, et estimées sous-elliptiques sur certains groupes nilpotents. *Acta Math.* **139**, 96-153.

[57] Goodman, R. (1977). Filtrations and asymptotic automorphisms on nilpotent Lie groups. *J. Differential Geom.* **12**, 183-196.

[58] Graczyk, P. (1994). Cramér theorem on symmetric spaces of noncompact type. *J. Theoret. Probab.* **7(3)**,609-613.

[59] Grenander, U. (1963). *Probabilities on algebraic structures.* Almquist & Wiksell, Stockholm.

[60] Guivarc'h, Y. (1976). Une loi des grands nombres pour les groupes de Lie. In: *Séminaire de Probabilités I (Univ. Rennes)*, Exp. No. 8.

[61] Hahn, M. G., Hudson, W. N., Veeh, J. A. (1989). Operator stable laws: Series representations and domains of normal attraction. *J. Theoret. Probab.* **2(1)**, 3-35.

[62] Häusler, E. (1988). *Laws of the iterated logarithm for sums of order statistics from a distribution with a regularly varying upper tail.* Habilitationsschrift, University of München.

[63] Hazod, W. (1971). Ueber Wurzeln und Logarithmen beschränkter Masse. *Z. Wahrscheinlichkeitstheorie verw. Geb.* **20**, 259-270.

[64] Hazod, W. (1977). *Stetige Faltungshalbgruppen von Wahrscheinlichkeitsmassen und erzeugende Distributionen.* Lecture Notes in Mathematics 595, Springer, Berlin.

[65] Hazod, W. (1982). Stable probabilities on locally compact groups. In: Heyer, H. (ed.). *Probability measures on groups.* Lecture Notes in Mathematics 928, Springer, Berlin, 183-211.

[66] Hazod, W. (1984a). Remarks on [semi-]stable probabilities. In: Heyer, H. (ed.). *Probability measures on groups VII.* Lecture Notes in Mathematics 1064, Springer, Berlin, 182-203.

[67] Hazod, W. (1984b). Stable and semistable probabilites on groups and on vector spaces. In: Szynal, D., Weron, A. (ed.). *Probability theory on vector spaces III.* Lecture Notes in Mathematics 1080. Springer, Berlin, 69-89.

[68] Hazod, W. (1986). Stable probability measures on groups and on vector spaces: A survey. In: Heyer, H. (ed.). *Probability measures on groups VIII.* Lecture Notes in Mathematics 1210. Springer, Berlin, 304-352.

[69] Hazod, W. (1993). Probabilities on contractible locally compact groups: The existence of universal distributions in the sense of W. Doeblin. *Ann. Inst. H. Poincaré Probab. Statist.* **29(3)**, 339-356

[70] Hazod, W., Nobel, S. (1989). Convergence-of-types theorem for simply connected nilpotent Lie groups. In: Heyer, H. (ed.). *Probability measures on groups IX.* Lecture Notes in Mathematics 1379. Springer, Berlin, 90-106.

[71] Hazod, W., Scheffler, H. P. (1993). The domains of partial attraction of probabilities on groups and on vector spaces. *J. Theoret. Probab.* **6(1)**, 175-186.

[72] Hazod, W., Siebert, E. (1986). Continuous automorphism groups on a locally compact group contracting modulo a compact subgroup and applications to stable semigroups. *Semigroup Forum* **33**, 111-143.

[73] Hazod, W., Siebert, E. (1988). Automorphisms on a Lie group contracting modulo a compact subgroup and applications to semistable convolution semigroups. *J. Theoret. Probab.* **1(2)**, 211-225.

[74] Hazod, W., Siebert, E. (1995). *Stable probability measures on euclidean spaces and on locally compact groups: Structural properties and limit theorems.* In preparation.

[75] Hebisch, W., Sikora, A. (1990). A smooth subadditive homogeneous norm on a homogeneous group. *Studia Math.* **96**, 231-236.

[76] Helffer, B. (1980). Hypoellipticité analytique sur des groupes nilpotents de rang 2 (d'après G. Métivier). *Séminaire Goulaouic-Schwartz 1979/80* 1.

[77] Helmes, K. (1986). The "local" law of the iterated logarithm for processes related to Lévy's stochastic area process. *Studia Math.* **LXXXIII**, 229-237.

[78] Helmes, K., Schwane, A. (1983). Lévy's stochastic area formula in higher dimensions. *J. Funct. Anal.* **54**, 177-192.

[79] Hewitt, E., Stromberg, K. (1965). *Real and abstract analysis.* Springer, New York.

[80] Heyde, C. C. (1967). On large deviation problems for sums of random variables which are not attracted to the normal law. *Ann. Math. Statist.* **38**, 1575-1578.

[81] Heyer, H. (1977). *Probability measures on locally compact groups.* Springer, Berlin.

[82] Heyer, H., Pap, G. (1996). *Convergence of noncommutative triangular arrays of probability measures on a Lie group.* Preprint.

[83] Howe, R. (1980). On the role of the Heisenberg group in harmonic analysis. *Bull. Amer. Math. Soc.* **3(2)**, 821-843.

[84] Hudson, W. N., Jurek, Z. J., Veeh, J. A. (1986). The symmetry group and exponents of operator stable probability measures. *Ann. Prob.* **14**, 1014-1023.

[85] Hudson, W. N., Mason, J. D. (1981). Exponents of operator-stable laws. In: Beck, A. (ed.). *Probability in Banach spaces III.* Lecture Notes in Mathematics 860. Springer, Berlin, 291-298.

[86] Hulanicki, A. (1976). The distribution of energy in the Brownian motion in the gaussian field and analytic hypoellipticity of certain subelliptic operators on the Heisenberg group. *Studia Math.* **LVI**, 165-173.

[87] Hunt, G. A. (1956). Semi-groups of measures on Lie groups. *Trans. Amer. Math. Soc.* **81**, 264-293.

[88] Hunt, G. A. (1958). Markov processes and potentials III. *Illinois J. Math.* **2**, 151-213.

[89] Ikeda, N., Watanabe, S. (1989). *Stochastic differential equations and diffusion processes.* North-Holland.

[90] Janson, S., Wichura, M. J. (1983). Invariance principles for stochastic area and related stochastic integrals. *Stochastic Process. Appl.* **16**, 71-84.

[91] Janssen, A. (1989). The domain of attraction of stable laws and extreme order statistics. *Prob. Math. Stat.* **10(2)**, 205-222.

[92] Jurek, Z. J. (1980). On stability of probability measures in euclidean spaces. In: Weron, A. (ed.). *Probability theory on vector spaces II*. Lecture Notes in Mathematics 828. Springer, Berlin, 129-145.

[93] Jurek, Z. J. (1984). Polar coordinates in Banach spaces. *Bull. Acad. Pol. Sci. Math.* **32**, 61-66.

[94] Kaplan, A. (1980). Fundamental solutions for a class of hypoelliptic PDE generated by composition of quadratic forms. *Trans. Amer. Math. Soc.* **258(1)**, 147-153.

[95] Kawada, Y., Ito, K. (1940). On the probability distribution on a compact group I. *Proc. Phys.-Math. Soc. Japan* **22**, 977-998.

[96] Kochen, S. B., Stone, C. J. (1964). A note on the Borel-Cantelli lemma. *Ill. J. of Math.* **8**, 248-251.

[97] Komlós, J. (1967). A generalisation of a problem of Steinhaus. *Acta Math. Acad. Sci. Hungar.* **18**, 217-229.

[98] Korányi, A. (1983). Geometric aspects of analysis on the Heisenberg group. In: De Michele, L., Ricci, F. (ed.). *Topics in modern harmonic analysis. Vol. I.* Istituto Nazionale di Alta Matematica Francesco Severi, Rome, 209-258.

[99] Korányi, A. (1985). Geometric properties of Heisenberg-type groups. *Adv. Math.* **56(1)**, 28-38.

[100] Kuelbs, J., Ledoux, M. (1986). Extreme values and a gaussian central limit theorem. *Probab. Th. Rel. Fields* **74**, 341-355.

[101] Kunita, H. (1994a). Convolution semigroups of stable distributions over a nilpotent Lie group. *Proc. Japan Acad. Ser. A* **70(10)**, 305-310.

[102] Kunita, H. (1994b). Stable Lévy processes on nilpotent Lie groupsp. In: *Stochastic analysis on infinite-dimensional spaces.* Pitman Research Notes in Math. 310.

[103] Kunita, H. (1995). Stable limit distributions over a nilpotent Lie group. *Proc. Japan Acad. Ser. A* **71(1)**, 1-5.

[104] Lamperti, J. (1963). Wiener's test and Markov chains. *J. Math. Anal. Appl.* **6**, 58-66.

[105] Le Page, R., Woodroofe, M. Zinn, J. (1981). Convergence to a stable distribution via order statistics. *Ann. Prob.* **9(4)**, 624-632.

[106] Léandre, R. (1988). Sur le théorème d'Atiyah-Singer. *Probab. Th. Rel. Fields* **80**, 119-137.

[107] Lévy, P. (1939). L'addition des variables aléatoires définies sur une circonférence. *Bull. Soc. Math. France* **67**, 1-41.

[108] Loève, M. (1977). *Probability theory I.* 4th ed. Springer, Berlin.

[109] Loève, M. (1978). *Probability theory II.* 4th ed. Springer, Berlin.

[110] Meerschaert, M. M. (1986). Domains of attraction of non-normal operator-stable laws. *J. Multivariate Anal.* **19**, 342-347.

[111] Métivier, G. (1980). Hypoellipticité analytique des groupes nilpotents de rang 2. *Duke Math. J.* **47**, 195-221.

[112] Molchanov, I. S. (1993). *Limit theorems for unions of random closed sets.* Lecture Notes in Mathematics 1561. Springer, Berlin.

[113] Neuenschwander, D. (1991). *Trimmed products on simply connected step 2-nilpotent Lie groups.* Dissertation, University of Bern.

[114] Neuenschwander, D. (1992). Limits of commutative triangular systems on simply connected step 2-nilpotent Lie groups. *J. Theoret. Probab.* **5(1)**, 217-222.

[115] Neuenschwander, D. (1993). Gauss measures in the sense of Bernstein on the Heisenberg group. *Prob. Math. Stat.* **14(2)**,253-256.

[116] Neuenschwander, D. (1995a). A gaussian central limit theorem for trimmed products on simply connected step 2-nilpotent Lie groups. *J. Theoret. Probab.* **8(1)**, 165-174.

[117] Neuenschwander, D. (1995b). Laws of large numbers on simply connected step 2-nilpotent Lie groups. To appear in: *Prob. Math. Stat.*

[118] Neuenschwander, D. (1995c). Domains of attraction of stable semigroups on simply connected nilpotent Lie groups. To appear in: Zolotarev, V. M. (ed.) *Stability problems for stochastic models.* Proceedings of the 1994 Eger conference. J. Math. Science.

[119] Neuenschwander, D. (1995d). Lightly trimmed products on simply connected step 2-nilpotent Lie groups. To appear in: Heyer, H. (ed.). *Probability measures on groups and related structures.* World Scientific.

[120] Neuenschwander, D. (1995e). Universal laws on simply connected step 2-nilpotent Lie groups. *Soviet Math. (Iz. VUZ)* **6(397)**,68-73.

[121] Neuenschwander, D. (1995f). Triangular systems on discrete subgroups of simply connected nilpotent Lie groups. *Publ. Math. Debrecen* **48(3-4)**, 1-5.

[122] Neuenschwander, D. (1995g). Commutative infinitesimal triangular systems on Euclidean motion groups. To appear in: *Statist. Probab. Lett.*

[123] Neuenschwander, D. (1995h). The Marcinkiewicz-Zygmund law of large numbers on the group of Euclidean motions and the diamond group. To appear in: Zolotarev, V. M. (ed.) *Stability problems for stochastic models.* Proceedings of the 1995 Kazan conference. J. Math. Science.

[124] Neuenschwander, D. (1995i). *Uniqueness of embedding into Poisson semigroups with boundedly supported Lévy measures on simply connected nilpotent Lie groups.* Preprint.

[125] Neuenschwander, D., Scheffler, H. P. (1995). The "Two series theorem" for symmetric random variables on nilpotent Lie groups. *Publ. Math. Debrecen* **47(1-2)**, 1-7.

[126] Neuenschwander, D., Scheffler, H. P. (1996). Laws of the iterated logarithm for the central part of (semi-)stable measures on the Heisenberg groups. *Monatsh. Math.* **121**, 265-274.

[127] Neuenschwander, D., Schott, R. (1995). On the local and asymptotic behavior of Brownian motion on simply connected nilpotent Lie groups. *J. Theoret. Probab.* **8(4)**, 795-806.

[128] Neuenschwander, D., Schott, R. (1996). *The Bernstein and Skitovič-Darmois characterization theorems for Gaussian distributions on groups, symmetric spaces, and quantum groups.* In preparation.

[129] Nguyen, V. T. (1981). A new version of Doeblin's theorem. *Ann. Inst. H. Poincaré Probab. Statist.* **XVII(2)**, 213-217.

[130] Nobel, S. (1988). *Grenzwertsätze für Wahrscheinlichkeitsmasse auf einfach zusammenhängenden nilpotenten Liegruppen.* Dissertation, University of Dortmund.

[131] Nobel, S. (1991). Limit theorems for probability measures on simply connected nilpotent Lie groups. *J. Theoret. Probab.* **4(2)**, 261-284.

[132] Ohring, P. (1993). A central limit theorem on Heisenberg type groups. II. *Proc. Amer. Math. Soc.* **118(4)**, 1313-1318.

[133] Pap, G. (1989). *Rate of convergence in CLT on Heisenberg group.* Unpublished manuscript.

[134] Pap, G. (1991a). Rate of convergence in CLT on stratified groups. *J. Multivariate Anal.* **38**, 333-365.

[135] Pap, G. (1991b). A new proof of the central limit theorem on stratified Lie groups. In: Heyer, H. (ed.). *Probability measures on groups X.* Plenum Press, New York, 329-336.

[136] Pap, G. (1992). *Lindeberg theorem on stratified nilpotent Lie groups.* Technical Report 92/31, Department of Mathematics, University of Debrecen.

[137] Pap, G. (1993). Central limit theorems on nilpotent Lie groups. *Prob. Math. Stat.* **14**, 287-312.

[138] Pap, G. (1994). Uniqueness of embedding into a gaussian semigroup on a nilpotent Lie group. *Arch. Math. (Basel)* **62**, 282-288.

[139] Pap, G. (1995). Edgeworth expansions in nilpotent Lie groups. Submitted to: Heyer, H. (ed.). *Probability measures on groups and related structures XI*. World Scientific.

[140] Pap, G. (1996a). *Functional central limit theorems and hemigroups of probability measures on a Lie group I*. Preprint.

[141] Pap, G. (1996b). *Functional central limit theorems and hemigroups of probability measures on a Lie group II*. Preprint.

[142] Port, S. C., Stone, C. J. (1970). Classical potential theory and Brownian motion. *Proceedings of the 6th Berkeley Symposium on Mathematical Statistics and Probabiliy Vol. III (Probability theory)*, 143-176.

[143] Pratelli. L. (1989). La loi des grands nombres pour une suite échangeable. In: Azéma, J. et al. (ed.). *Séminaire de Probabilités XXIII*. Lecture Notes in Mathematics 1372. Springer, Berlin.

[144] Raugi, A. (1978). Théorème de la limite centrale sur les groupes nilpotents. *Z. Wahrscheinlichkeitstheorie verw. Geb.* **43**, 149-172.

[145] Rémillard, B. (1994). On Chung's law of the iterated logarithm for some stochastic integrals. *Ann. Prob.* **22(4)**, 1794 1802.

[146] Riddhi Shah (1991). Infinitely divisible measures on p-adic groups. *J. Theoret. Probab.* **4**, 391-405.

[147] Riddhi Shah (1995). Convergence of types theorems on p-adic algebraic groups. In: Heyer, H. (ed.) *Probability measures on groups and related structures*. World Scientific, 357-363.

[148] Rogers, C. A., Taylor, S. J. (1961). Functions continuous and singular with respect to a Hausdorff measure. *Mathematika* **8**, 1-31.

[149] Roynette, B. (1975). Croissance et mouvements Browniens d'un groupe de Lie nilpotent et simplement connexe. *Z. Wahrscheinlichkeitstheorie verw. Geb. 32*, 133-138.

[150] Ruzsa, I. Z. (1988). Infinite divisibility II. *J. Theoret. Probab.* **1**, 327-339.

[151] Rvačeva, E. L. (1962). On domains of attraction of multi-dimensional distributions. *Sel. Transl. Math. Stat. Prob.* **2**, 183-205.

[152] Scheffler, H. P. (1992). *Anziehungsbereiche stabiler Wahrscheinlichkeitsmasse auf stratifizierbaren Lie-Gruppen.* Dissertation, University of Dortmund.

[153] Scheffler, H. P. (1993). Domains of attraction of stable measures on the Heisenberg group. *Prob. Math. Stat.* **14**, 327-345.

[154] Scheffler, H. P. (1994). D-domains of attraction of stable measures on stratified Lie groups. *J. Theoret. Probab.* **7(4)**, 767-792.

[155] Scheffler, H. P. (1995a). A law of the iterated logarithm for stable laws on homogeneous groups. *Publ. Math. Debrecen* **47(3-4)**, 377-391.

[156] Scheffler, H. P. (1995b). *A law of the iterated logarithm for semistable laws on vector spaces and homogeneous groups.* Preprint.

[157] Schempp, W. (1988). *Harmonic analysis on the Heisenberg nilpotent Lie group.* Longman, Harlow.

[158] Schott, R. (1981). Une loi de logarithme itéré pour certaines intégrales stochastiques. *C. R. Acad. Sci. Paris I*, 295-298.

[159] Serre, J.-P. (1965). *Lie algebras and Lie groups.* Benjamin, New York.

[160] Sharpe, M. (1969). Operator stable probability distributions on vector groups. *Trans. Amer. Math. Soc.* **136**, 51-65.

[161] Shi, Z. (1995). Liminf behaviours of the windings and Lévy's stochastic areas of planar Brownian motion. In: Azéma, J., Meyer, P. A., Yor, M. (ed.). *Séminaire de Probabilités XXVIII.* Lecture Notes in Mathematics 1583. Springer, Berlin, 122-137.

[162] Shiryayev, A. N. (1984). *Probability.* Springer, Berlin.

[163] Siebert, E. (1973). Ueber die Erzeugung von Faltungshalbgruppen auf beliebigen lokalkompakten Gruppen. *Math. Z.* **131**, 313-333.

[164] Siebert, E. (1981). Fourier analysis and limit theorems for convolution semigroups on a locally compact group. *Adv. Math.* **39**, 111-154.

[165] Siebert, E. (1982). Continuous hemigroups of probability measures on a Lie group. In: Heyer, H. (ed.). *Probability measures on groups.* Lecture Notes in Mathematics 928. Springer, Berlin, 362-402.

[166] Spitzer, F. (1973). Discussion on Professor Kingman's paper: Subadditive ergodic theory. *Ann. Prob.* **1(6)**, 904-905.

[167] Stroock, D. W., Varadhan, S. R. S. (1973). Limit theorems for random walks on Lie groups. *Sankhya Ser. A* **35**, 277-294.

[168] Sznitman, A. S. (1987). Some bounds and limiting results for the measure of Wiener sausage of small radius associated with elliptic diffusions. *Stoch. Processes Appl.* **25**, 1-25.

[169] Taylor, M. E. (1986). *Noncommutative harmonic analysis*. American Mathematical Society, Providence.

[170] Telöken, K. (1996). *Grenzwertsätze für Wahrscheinlichkeitsmasse auf total unzusammenhängenden Gruppen*. Dissertation, University of Dortmund.

[171] Tutubalin, V. N. (1964). Compositions of measures on the simplest nilpotent group. *Th. Prob. Appl.* **9**, 479-487.

[172] Tutubalin, V. N. (1969). Some theorems of the type of the strong law of large numbers. *Th. Probab. App.* **14**, 319 326.

[173] Watkins, J. (1989). Donsker's invariance principle for Lie groups. *Ann. Prob.* **17(3)**, 1220-1242.

[174] Wehn, D. F. (1962). Probabilities on Lie groups. *Proc. Nat. Acad. Sci. USA* **48**, 791-795.

[175] von Weizsäcker, H., Winkler, G. (1990). *Stochastic integrals*. Vieweg, Braunschweig.

[176] Yor, M. (1991). The laws of some Brownian functionals. *Proceedings of the international congress of mathematicians, Kyoto, 1990*. Springer, Berlin, 1105-1112.

Index

Lecture Notes in Mathematics

For information about Vols. 1–1449
please contact your bookseller or Springer-Verlag

Vol. 1450: H. Fujita, T. Ikebe, S.T. Kuroda (Eds.), Functional-Analytic Methods for Partial Differential Equations. Proceedings, 1989. VII, 252 pages. 1990.

Vol. 1451: L. Alvarez-Gaumé, E. Arbarello, C. De Concini, N.J. Hitchin, Global Geometry and Mathematical Physics. Montecatini Terme 1988. Seminar. Editors: M. Francaviglia, F. Gherardelli. IX, 197 pages. 1990.

Vol. 1452: E. Hlawka, R.F. Tichy (Eds.), Number-Theoretic Analysis. Seminar, 1988–89. V, 220 pages. 1990.

Vol. 1453: Yu.G. Borisovich, Yu.E. Gliklikh (Eds.), Global Analysis – Studies and Applications IV. V, 320 pages. 1990.

Vol. 1454: F. Baldassari, S. Bosch, B. Dwork (Eds.), p-adic Analysis. Proceedings, 1989. V, 382 pages. 1990.

Vol. 1455: J.-P. Françoise, R. Roussarie (Eds.), Bifurcations of Planar Vector Fields. Proceedings, 1989. VI, 396 pages. 1990.

Vol. 1456: L.G. Kovács (Ed.), Groups – Canberra 1989. Proceedings. XII, 198 pages. 1990.

Vol. 1457: O. Axelsson, L.Yu. Kolotilina (Eds.), Preconditioned Conjugate Gradient Methods. Proceedings, 1989. V, 196 pages. 1990.

Vol. 1458: R. Schaaf, Global Solution Branches of Two Point Boundary Value Problems. XIX, 141 pages. 1990.

Vol. 1459: D. Tiba, Optimal Control of Nonsmooth Distributed Parameter Systems. VII, 159 pages. 1990.

Vol. 1460: G. Toscani, V. Boffi, S. Rionero (Eds.), Mathematical Aspects of Fluid Plasma Dynamics. Proceedings, 1988. V, 221 pages. 1991.

Vol. 1461: R. Gorenflo, S. Vessella, Abel Integral Equations. VII, 215 pages. 1991.

Vol. 1462: D. Mond, J. Montaldi (Eds.), Singularity Theory and its Applications. Warwick 1989, Part I. VIII, 405 pages. 1991.

Vol. 1463: R. Roberts, I. Stewart (Eds.), Singularity Theory and its Applications. Warwick 1989, Part II. VIII, 322 pages. 1991.

Vol. 1464: D. L. Burkholder, E. Pardoux, A. Sznitman, Ecole d'Eté de Probabilités de Saint- Flour XIX-1989. Editor: P. L. Hennequin. VI, 256 pages. 1991.

Vol. 1465: G. David, Wavelets and Singular Integrals on Curves and Surfaces. X, 107 pages. 1991.

Vol. 1466: W. Banaszczyk, Additive Subgroups of Topological Vector Spaces. VII, 178 pages. 1991.

Vol. 1467: W. M. Schmidt, Diophantine Approximations and Diophantine Equations. VIII, 217 pages. 1991.

Vol. 1468: J. Noguchi, T. Ohsawa (Eds.), Prospects in Complex Geometry. Proceedings, 1989. VII, 421 pages. 1991.

Vol. 1469: J. Lindenstrauss, V. D. Milman (Eds.), Geometric Aspects of Functional Analysis. Seminar 1989-90. XI, 191 pages. 1991.

Vol. 1470: E. Odell, H. Rosenthal (Eds.), Functional Analysis. Proceedings, 1987-89. VII, 199 pages. 1991.

Vol. 1471: A. A. Panchishkin, Non-Archimedean L-Functions of Siegel and Hilbert Modular Forms. VII, 157 pages. 1991.

Vol. 1472: T. T. Nielsen, Bose Algebras: The Complex and Real Wave Representations. V, 132 pages. 1991.

Vol. 1473: Y. Hino, S. Murakami, T. Naito, Functional Differential Equations with Infinite Delay. X, 317 pages. 1991.

Vol. 1474: S. Jackowski, B. Oliver, K. Pawałowski (Eds.), Algebraic Topology, Poznań 1989. Proceedings. VIII, 397 pages. 1991.

Vol. 1475: S. Busenberg, M. Martelli (Eds.), Delay Differential Equations and Dynamical Systems. Proceedings, 1990. VIII, 249 pages. 1991.

Vol. 1476: M. Bekkali, Topics in Set Theory. VII, 120 pages. 1991.

Vol. 1477: R. Jajte, Strong Limit Theorems in Noncommutative L_2-Spaces. X, 113 pages. 1991.

Vol. 1478: M.-P. Malliavin (Ed.), Topics in Invariant Theory. Seminar 1989-1990. VI, 272 pages. 1991.

Vol. 1479: S. Bloch, I. Dolgachev, W. Fulton (Eds.), Algebraic Geometry. Proceedings, 1989. VII, 300 pages. 1991.

Vol. 1480: F. Dumortier, R. Roussarie, J. Sotomayor, H. Zoladek, Bifurcations of Planar Vector Fields: Nilpotent Singularities and Abelian Integrals. VIII, 226 pages. 1991.

Vol. 1481: D. Ferus, U. Pinkall, U. Simon, B. Wegner (Eds.), Global Differential Geometry and Global Analysis. Proceedings, 1991. VIII, 283 pages. 1991.

Vol. 1482: J. Chabrowski, The Dirichlet Problem with L^2-Boundary Data for Elliptic Linear Equations. VI, 173 pages. 1991.

Vol. 1483: E. Reithmeier, Periodic Solutions of Nonlinear Dynamical Systems. VI, 171 pages. 1991.

Vol. 1484: H. Delfs, Homology of Locally Semialgebraic Spaces. IX, 136 pages. 1991.

Vol. 1485: J. Azéma, P. A. Meyer, M. Yor (Eds.), Séminaire de Probabilités XXV. VIII, 440 pages. 1991.

Vol. 1486: L. Arnold, H. Crauel, J.-P. Eckmann (Eds.), Lyapunov Exponents. Proceedings, 1990. VIII, 365 pages. 1991.

Vol. 1487: E. Freitag, Singular Modular Forms and Theta Relations. VI, 172 pages. 1991.

Vol. 1488: A. Carboni, M. C. Pedicchio, G. Rosolini (Eds.), Category Theory. Proceedings, 1990. VII, 494 pages. 1991.

Vol. 1489: A. Mielke, Hamiltonian and Lagrangian Flows on Center Manifolds. X, 140 pages. 1991.

Vol. 1490: K. Metsch, Linear Spaces with Few Lines. XIII, 196 pages. 1991.

Vol. 1491: E. Lluis-Puebla, J.-L. Loday, H. Gillet, C. Soulé, V. Snaith, Higher Algebraic K-Theory: an overview. IX, 164 pages. 1992.

Vol. 1492: K. R. Wicks, Fractals and Hyperspaces. VIII, 168 pages. 1991.

Vol. 1493: E. Benoît (Ed.), Dynamic Bifurcations. Proceedings, Luminy 1990. VII, 219 pages. 1991.

Vol. 1494: M.-T. Cheng, X.-W. Zhou, D.-G. Deng (Eds.), Harmonic Analysis. Proceedings, 1988. IX, 226 pages. 1991.

Vol. 1495: J. M. Bony, G. Grubb, L. Hörmander, H. Komatsu, J. Sjöstrand, Microlocal Analysis and Applications. Montecatini Terme, 1989. Editors: L. Cattabriga, L. Rodino. VII, 349 pages. 1991.

Vol. 1496: C. Foias, B. Francis, J. W. Helton, H. Kwakernaak, J. B. Pearson, H_∞-Control Theory. Como, 1990. Editors: E. Mosca, L. Pandolfi. VII, 336 pages. 1991.

Vol. 1497: G. T. Herman, A. K. Louis, F. Natterer (Eds.), Mathematical Methods in Tomography. Proceedings 1990. X, 268 pages. 1991.

Vol. 1498: R. Lang, Spectral Theory of Random Schrödinger Operators. X, 125 pages. 1991.

Vol. 1499: K. Taira, Boundary Value Problems and Markov Processes. IX, 132 pages. 1991.

Vol. 1500: J.-P. Serre, Lie Algebras and Lie Groups. VII, 168 pages. 1992.

Vol. 1501: A. De Masi, E. Presutti, Mathematical Methods for Hydrodynamic Limits. IX, 196 pages. 1991.

Vol. 1502: C. Simpson, Asymptotic Behavior of Monodromy. V, 139 pages. 1991.

Vol. 1503: S. Shokranian, The Selberg-Arthur Trace Formula (Lectures by J. Arthur). VII, 97 pages. 1991.

Vol. 1504: J. Cheeger, M. Gromov, C. Okonek, P. Pansu, Geometric Topology: Recent Developments. Editors: P. de Bartolomeis, F. Tricerri. VII, 197 pages. 1991.

Vol. 1505: K. Kajitani, T. Nishitani, The Hyperbolic Cauchy Problem. VII, 168 pages. 1991.

Vol. 1506: A. Buium, Differential Algebraic Groups of Finite Dimension. XV, 145 pages. 1992.

Vol. 1507: K. Hulek, T. Peternell, M. Schneider, F.-O. Schreyer (Eds.), Complex Algebraic Varieties. Proceedings, 1990. VII, 179 pages. 1992.

Vol. 1508: M. Vuorinen (Ed.), Quasiconformal Space Mappings. A Collection of Surveys 1960-1990. IX, 148 pages. 1992.

Vol. 1509: J. Aguadé, M. Castellet, F. R. Cohen (Eds.), Algebraic Topology - Homotopy and Group Cohomology. Proceedings, 1990. X, 330 pages. 1992.

Vol. 1510: P. P. Kulish (Ed.), Quantum Groups. Proceedings, 1990. XII, 398 pages. 1992.

Vol. 1511: B. S. Yadav, D. Singh (Eds.), Functional Analysis and Operator Theory. Proceedings, 1990. VIII, 223 pages. 1992.

Vol. 1512: L. M. Adleman, M.-D. A. Huang, Primality Testing and Abelian Varieties Over Finite Fields. VII, 142 pages. 1992.

Vol. 1513: L. S. Block, W. A. Coppel, Dynamics in One Dimension. VIII, 249 pages. 1992.

Vol. 1514: U. Krengel, K. Richter, V. Warstat (Eds.), Ergodic Theory and Related Topics III. Proceedings, 1990. VIII, 236 pages. 1992.

Vol. 1515: E. Ballico, F. Catanese, C. Ciliberto (Eds.), Classification of Irregular Varieties. Proceedings, 1990. VII, 149 pages. 1992.

Vol. 1516: R. A. Lorentz, Multivariate Birkhoff Interpolation. IX, 192 pages. 1992.

Vol. 1517: K. Keimel, W. Roth, Ordered Cones and Approximation. VI, 134 pages. 1992.

Vol. 1518: H. Stichtenoth, M. A. Tsfasman (Eds.), Coding Theory and Algebraic Geometry. Proceedings, 1991. VIII, 223 pages. 1992.

Vol. 1519: M. W. Short, The Primitive Soluble Permutation Groups of Degree less than 256. IX, 145 pages. 1992.

Vol. 1520: Yu. G. Borisovich, Yu. E. Gliklikh (Eds.), Global Analysis – Studies and Applications V. VII, 284 pages. 1992.

Vol. 1521: S. Busenberg, B. Forte, H. K. Kuiken, Mathematical Modelling of Industrial Process. Bari, 1990. Editors: V. Capasso, A. Fasano. VII, 162 pages. 1992.

Vol. 1522: J.-M. Delort, F. B. I. Transformation. VII, 101 pages. 1992.

Vol. 1523: W. Xue, Rings with Morita Duality. X, 168 pages. 1992.

Vol. 1524: M. Coste, L. Mahé, M.-F. Roy (Eds.), Real Algebraic Geometry. Proceedings, 1991. VIII, 418 pages. 1992.

Vol. 1525: C. Casacuberta, M. Castellet (Eds.), Mathematical Research Today and Tomorrow. VII, 112 pages. 1992.

Vol. 1526: J. Azéma, P. A. Meyer, M. Yor (Eds.), Séminaire de Probabilités XXVI. X, 633 pages. 1992.

Vol. 1527: M. I. Freidlin, J.-F. Le Gall, Ecole d'Eté de Probabilités de Saint-Flour XX – 1990. Editor: P. L. Hennequin. VIII, 244 pages. 1992.

Vol. 1528: G. Isac, Complementarity Problems. VI, 297 pages. 1992.

Vol. 1529: J. van Neerven, The Adjoint of a Semigroup of Linear Operators. X, 195 pages. 1992.

Vol. 1530: J. G. Heywood, K. Masuda, R. Rautmann, S. A. Solonnikov (Eds.), The Navier-Stokes Equations II – Theory and Numerical Methods. IX, 322 pages. 1992.

Vol. 1531: M. Stoer, Design of Survivable Networks. IV, 206 pages. 1992.

Vol. 1532: J. F. Colombeau, Multiplication of Distributions. X, 184 pages. 1992.

Vol. 1533: P. Jipsen, H. Rose, Varieties of Lattices. X, 162 pages. 1992.

Vol. 1534: C. Greither, Cyclic Galois Extensions of Commutative Rings. X, 145 pages. 1992.

Vol. 1535: A. B. Evans, Orthomorphism Graphs of Groups. VIII, 114 pages. 1992.

Vol. 1536: M. K. Kwong, A. Zettl, Norm Inequalities for Derivatives and Differences. VII, 150 pages. 1992.

Vol. 1537: P. Fitzpatrick, M. Martelli, J. Mawhin, R. Nussbaum, Topological Methods for Ordinary Differential Equations. Montecatini Terme, 1991. Editors: M. Furi, P. Zecca. VII, 218 pages. 1993.

Vol. 1538: P.-A. Meyer, Quantum Probability for Probabilists. X, 287 pages. 1993.

Vol. 1539: M. Coornaert, A. Papadopoulos, Symbolic Dynamics and Hyperbolic Groups. VIII, 138 pages. 1993.

Vol. 1540: H. Komatsu (Ed.), Functional Analysis and Related Topics, 1991. Proceedings. XXI, 413 pages. 1993.

Vol. 1541: D. A. Dawson, B. Maisonneuve, J. Spencer, Ecole d' Eté de Probabilités de Saint-Flour XXI - 1991. Editor: P. L. Hennequin. VIII, 356 pages. 1993.

Vol. 1542: J. Fröhlich, Th. Kerler, Quantum Groups, Quantum Categories and Quantum Field Theory. VII, 431 pages. 1993.

Vol. 1543: A. L. Dontchev, T. Zolezzi, Well-Posed Optimization Problems. XII, 421 pages. 1993.

Vol. 1544: M. Schürmann, White Noise on Bialgebras. VII, 146 pages. 1993.

Vol. 1545: J. Morgan, K. O'Grady, Differential Topology of Complex Surfaces. VIII, 224 pages. 1993.

Vol. 1546: V. V. Kalashnikov, V. M. Zolotarev (Eds.), Stability Problems for Stochastic Models. Proceedings, 1991. VIII, 229 pages. 1993.

Vol. 1547: P. Harmand, D. Werner, W. Werner, M-ideals in Banach Spaces and Banach Algebras. VIII, 387 pages. 1993.

Vol. 1548: T. Urabe, Dynkin Graphs and Quadrilateral Singularities. VI, 233 pages. 1993.

Vol. 1549: G. Vainikko, Multidimensional Weakly Singular Integral Equations. XI, 159 pages. 1993.

Vol. 1550: A. A. Gonchar, E. B. Saff (Eds.), Methods of Approximation Theory in Complex Analysis and Mathematical Physics IV, 222 pages, 1993.

Vol. 1551: L. Arkeryd, P. L. Lions, P.A. Markowich, S.R. S. Varadhan. Nonequilibrium Problems in Many-Particle Systems. Montecatini, 1992. Editors: C. Cercignani, M. Pulvirenti. VII, 158 pages 1993.

Vol. 1552: J. Hilgert, K.-H. Neeb, Lie Semigroups and their Applications. XII, 315 pages. 1993.

Vol. 1553: J.-L- Colliot-Thélène, J. Kato, P. Vojta. Arithmetic Algebraic Geometry. Trento, 1991. Editor: E. Ballico. VII, 223 pages. 1993.

Vol. 1554: A. K. Lenstra, H. W. Lenstra, Jr. (Eds.), The Development of the Number Field Sieve. VIII, 131 pages. 1993.

Vol. 1555: O. Liess, Conical Refraction and Higher Microlocalization. X, 389 pages. 1993.

Vol. 1556: S. B. Kuksin, Nearly Integrable Infinite-Dimensional Hamiltonian Systems. XXVII, 101 pages. 1993.

Vol. 1557: J. Azéma, P. A. Meyer, M. Yor (Eds.), Séminaire de Probabilités XXVII. VI, 327 pages. 1993.

Vol. 1558: T. J. Bridges, J. E. Furter, Singularity Theory and Equivariant Symplectic Maps. VI, 226 pages. 1993.

Vol. 1559: V. G. Sprindžuk, Classical Diophantine Equations. XII, 228 pages. 1993.

Vol. 1560: T. Bartsch, Topological Methods for Variational Problems with Symmetries. X, 152 pages. 1993.

Vol. 1561: I. S. Molchanov, Limit Theorems for Unions of Random Closed Sets. X, 157 pages. 1993.

Vol. 1562: G. Harder, Eisensteinkohomologie und die Konstruktion gemischter Motive. XX, 184 pages. 1993.

Vol. 1563: E. Fabes, M. Fukushima, L. Gross, C. Kenig, M. Röckner, D. W. Stroock, Dirichlet Forms. Varenna, 1992. Editors: G. Dell'Antonio, U. Mosco. VII, 245 pages. 1993.

Vol. 1564: J. Jorgenson, S. Lang, Basic Analysis of Regularized Series and Products. IX, 122 pages. 1993.

Vol. 1565: L. Boutet de Monvel, C. De Concini, C. Procesi, P. Schapira, M. Vergne. D-modules, Representation Theory, and Quantum Groups. Venezia, 1992. Editors: G. Zampieri, A. D'Agnolo. VII, 217 pages. 1993.

Vol. 1566: B. Edixhoven, J.-H. Evertse (Eds.), Diophantine Approximation and Abelian Varieties. XIII, 127 pages. 1993.

Vol. 1567: R. L. Dobrushin, S. Kusuoka, Statistical Mechanics and Fractals. VII, 98 pages. 1993.

Vol. 1568: F. Weisz, Martingale Hardy Spaces and their Application in Fourier Analysis. VIII, 217 pages. 1994.

Vol. 1569: V. Totik, Weighted Approximation with Varying Weight. VI, 117 pages. 1994.

Vol. 1570: R. deLaubenfels, Existence Families, Functional Calculi and Evolution Equations. XV, 234 pages. 1994.

Vol. 1571: S. Yu. Pilyugin, The Space of Dynamical Systems with the C^0-Topology. X, 188 pages. 1994.

Vol. 1572: L. Göttsche, Hilbert Schemes of Zero-Dimensional Subschemes of Smooth Varieties. IX, 196 pages. 1994.

Vol. 1573: V. P. Havin, N. K. Nikolski (Eds.), Linear and Complex Analysis – Problem Book 3 – Part I. XXII, 489 pages. 1994.

Vol. 1574: V. P. Havin, N. K. Nikolski (Eds.), Linear and Complex Analysis – Problem Book 3 – Part II. XXII, 507 pages. 1994.

Vol. 1575: M. Mitrea, Clifford Wavelets, Singular Integrals, and Hardy Spaces. XI, 116 pages. 1994.

Vol. 1576: K. Kitahara, Spaces of Approximating Functions with Haar-Like Conditions. X, 110 pages. 1994.

Vol. 1577: N. Obata, White Noise Calculus and Fock Space. X, 183 pages. 1994.

Vol. 1578: J. Bernstein, V. Lunts, Equivariant Sheaves and Functors. V, 139 pages. 1994.

Vol. 1579: N. Kazamaki, Continuous Exponential Martingales and *BMO*. VII, 91 pages. 1994.

Vol. 1580: M. Milman, Extrapolation and Optimal Decompositions with Applications to Analysis. XI, 161 pages. 1994.

Vol. 1581: D. Bakry, R. D. Gill, S. A. Molchanov, Lectures on Probability Theory. Editor: P. Bernard. VIII, 420 pages. 1994.

Vol. 1582: W. Balser, From Divergent Power Series to Analytic Functions. X, 108 pages. 1994.

Vol. 1583: J. Azéma, P. A. Meyer, M. Yor (Eds.), Séminaire de Probabilités XXVIII. VI, 334 pages. 1994.

Vol. 1584: M. Brokate, N. Kenmochi, I. Müller, J. F. Rodriguez, C. Verdi, Phase Transitions and Hysteresis. Montecatini Terme, 1993. Editor: A. Visintin. VII, 291 pages. 1994.

Vol. 1585: G. Frey (Ed.), On Artin's Conjecture for Odd 2-dimensional Representations. VIII, 148 pages. 1994.

Vol. 1586: R. Nillsen, Difference Spaces and Invariant Linear Forms. XII, 186 pages. 1994.

Vol. 1587: N. Xi, Representations of Affine Hecke Algebras. VIII, 137 pages. 1994.

Vol. 1588: C. Scheiderer, Real and Étale Cohomology. XXIV, 273 pages. 1994.

Vol. 1589: J. Bellissard, M. Degli Esposti, G. Forni, S. Graffi, S. Isola, J. N. Mather, Transition to Chaos in Classical and Quantum Mechanics. Montecatini Terme, 1991. Editor: S. Graffi. VII, 192 pages. 1994.

Vol. 1590: P. M. Soardi, Potential Theory on Infinite Networks. VIII, 187 pages. 1994.

Vol. 1591: M. Abate, G. Patrizio, Finsler Metrics – A Global Approach. IX, 180 pages. 1994.

Vol. 1592: K. W. Breitung, Asymptotic Approximations for Probability Integrals. IX, 146 pages. 1994.

Vol. 1593: J. Jorgenson & S. Lang, D. Goldfeld, Explicit Formulas for Regularized Products and Series. VIII, 154 pages. 1994.

Vol. 1594: M. Green, J. Murre, C. Voisin, Algebraic Cycles and Hodge Theory. Torino, 1993. Editors: A. Albano, F. Bardelli. VII, 275 pages. 1994.

Vol. 1595: R.D.M. Accola, Topics in the Theory of Riemann Surfaces. IX, 105 pages. 1994.

Vol. 1596: L. Heindorf, L. B. Shapiro, Nearly Projective Boolean Algebras. X, 202 pages. 1994.

Vol. 1597: B. Herzog, Kodaira-Spencer Maps in Local Algebra. XVII, 176 pages. 1994.

Vol. 1598: J. Berndt, F. Tricerri, L. Vanhecke, Generalized Heisenberg Groups and Damek-Ricci Harmonic Spaces. VIII, 125 pages. 1995.

Vol. 1599: K. Johannson, Topology and Combinatorics of 3-Manifolds. XVIII, 446 pages. 1995.

Vol. 1600: W. Narkiewicz, Polynomial Mappings. VII, 130 pages. 1995.

Vol. 1601: A. Pott, Finite Geometry and Character Theory. VII, 181 pages. 1995.

Vol. 1602: J. Winkelmann, The Classification of Three-dimensional Homogeneous Complex Manifolds. XI, 230 pages. 1995.

Vol. 1603: V. Ene, Real Functions – Current Topics. XIII, 310 pages. 1995.

Vol. 1604: A. Huber, Mixed Motives and their Realization in Derived Categories. XV, 207 pages. 1995.

Vol. 1605: L. B. Wahlbin, Superconvergence in Galerkin Finite Element Methods. XI, 166 pages. 1995.

Vol. 1606: P.-D. Liu, M. Qian, Smooth Ergodic Theory of Random Dynamical Systems. XI, 221 pages. 1995.

Vol. 1607: G. Schwarz, Hodge Decomposition – A Method for Solving Boundary Value Problems. VII, 155 pages. 1995.

Vol. 1608: P. Biane, R. Durrett, Lectures on Probability Theory. VII, 210 pages. 1995.

Vol. 1609: L. Arnold, C. Jones, K. Mischaikow, G. Raugel, Dynamical Systems. Montecatini Terme, 1994. Editor: R. Johnson. VIII, 329 pages. 1995.

Vol. 1610: A. S. Üstunel, An Introduction to Analysis on Wiener Space. X, 95 pages. 1995.

Vol. 1611: N. Knarr, Translation Planes. VI, 112 pages. 1995.

Vol. 1612: W. Kühnel, Tight Polyhedral Submanifolds and Tight Triangulations. VII, 122 pages. 1995.

Vol. 1613: J. Azéma, M. Emery, P. A. Meyer, M. Yor (Eds.), Séminaire de Probabilités XXIX. VI, 326 pages. 1995.

Vol. 1614: A. Koshelev, Regularity Problem for Quasilinear Elliptic and Parabolic Systems. XXI, 255 pages. 1995.

Vol. 1615: D. B. Massey, Lê Cycles and Hypersurface Singularities. XI, 131 pages. 1995.

Vol. 1616: I. Moerdijk, Classifying Spaces and Classifying Topoi. VII, 94 pages. 1995.

Vol. 1617: V. Yurinsky, Sums and Gaussian Vectors. XI, 305 pages. 1995.

Vol. 1618: G. Pisier, Similarity Problems and Completely Bounded Maps. VII, 156 pages. 1996.

Vol. 1619: E. Landvogt, A Compactification of the Bruhat-Tits Building. VII, 152 pages. 1996.

Vol. 1620: R. Donagi, B. Dubrovin, E. Frenkel, E. Previato, Integrable Systems and Quantum Groups. VIII, 488 pages. 1996.

Vol. 1621: H. Bass, M. V. Otero-Espinar, D. N. Rockmore, C. P. L. Tresser, Cyclic Renormalization and Automorphism Groups of Rooted Trees. XXI, 136 pages. 1996.

Vol. 1622: E. D. Farjoun, Cellular Spaces, Null Spaces and Homotopy Localization. XIV, 199 pages. 1996.

Vol. 1623: H.P. Yap, Total Colourings of Graphs. VIII, 131 pages. 1996.

Vol. 1624: V. Brînzănescu, Holomorphic Vector Bundles over Compact Complex Surfaces. X, 170 pages. 1996.

Vol. 1625: S. Lang, Topics in the Cohomology of Groups. VII, 226 pages. 1996.

Vol. 1626: J. Azéma, M. Emery, M. Yor (Eds.), Séminaire de Probabilités XXX. VIII, 382 pages. 1996.

Vol. 1627: C. Graham, Th. G. Kurtz, S. Méléard, Ph. E. Protter, M. Pulvirenti, D. Talay, Probabilistic Models for Nonlinear Partial Differential Equations. X, 301 pages. 1996.

Vol. 1628: P.-H. Zieschang, An Algebraic Approach to Association Schemes. XII, 189 pages. 1996.

Vol. 1630: D. Neuenschwander, Probabilities on the Heisenberg Group: Limit Theorems and Brownian Motion. VIII, 139 pages. 1996.